Solutions Manual to Accompany

CHEMICAL THERMODYNAMICS

CHARLES E. REID

University of Florida

McGraw-Hill Publishing Company

New York St. Louis San Francisco Auckland Bogotá Caracas
Hamburg Lisbon London Madrid Mexico Milan
Montreal New Delhi Oklahoma City Paris San Juan
São Paulo Singapore Sydney Tokyo Toronto

Solutions Manual to Accompany
CHEMICAL THERMODYNAMICS

ISBN 0-07-051770-3

1 2 3 4 5 6 7 8 9 0 WHT WHT 8 9 3 2 1 0

INTRODUCTION

This manual contains worked-out solutions of all exercise problems in Chemical Thermodynamics, by C. E. Reid, published by McGraw-Hill Book Company in 1990. The following points should be noted:

Equations indicated by a complete reference, such as "Eq. (6.21)", refer to equations in the textbook; those referred to simply as (1), etc., are equations in the solution to the same problem.

In almost all cases intermediate results were retained in the storage registers of the calculator and recalled for use unchanged; only the answers were rounded. Those who make a practice of rounding intermediate results may get slightly different answers from those given here.

Rounding may not always be consistent. The practice, often taught to beginning students, to round to a fixed number of significant figures, is crude at best, since three significant figures (for example) beginning with 9 imply much higher precision than three beginning with 1. Calculating expected error bounds would be a much more desirable procedure, but the data for doing this are often not available. Moreover, if this were done with all problems, the time required would detract from the primary goal of learning thermodynamics.

The author would appreciate learning of any errors.

Charles E. Reid
Chemsitry Department
University of Florida
Gainesville, FL 32611

CHAPTER 1

1.1

$$\left(\frac{\partial A}{\partial b}\right)_h = \left(\frac{\partial A}{\partial b}\right)_s + \left(\frac{\partial A}{\partial s}\right)_b\left(\frac{\partial s}{\partial b}\right)_h$$

From Eq. (1.1) $(\partial A/\partial b)_h = h$

From Eq. (1.2) $\left(\frac{\partial A}{\partial b}\right)_s = (s^2 - b^2)^{1/2} - \frac{b^2}{(s^2 - b^2)^{1/2}} = \frac{s^2 - 2b^2}{(s^2 - b^2)^{1/2}}$

From Eq. (1.2) $(\partial A/\partial s)_b = bs/(s^2 - b^2)^{1/2}$

From $s^2 = b^2 + h^2$ we get $(\partial s/\partial b)_h = b/s$

Substituting these into the r.h.s. gives

$$\frac{s^2 - 2b^2}{(s^2 - b^2)^{1/2}} + \frac{bs}{(s^2 - b^2)^{1/2}} \cdot \frac{b}{s} = \frac{s^2 - b^2}{(s^2 - b^2)^{1/2}} = (s^2 - b^2)^{1/2} = h$$

in agreement with the l.h.s.

1.2

$$\left(\frac{\partial A}{\partial s}\right)_t = \left(\frac{\partial A}{\partial s}\right)_b + \left(\frac{\partial A}{\partial b}\right)_s\left(\frac{\partial b}{\partial s}\right)_t$$

From Eq. 1.4 we get $(\partial A/\partial s)_t = t$

From Eqs. 1.2 and 1.3 $b(s^2 - b^2)^{1/2} = b^2t/(b^2 - t^2)^{1/2}$

Squaring and dividing by b^2 gives $s^2 - b^2 = b^2t^2/(b^2 - t^2)$, from which $b^4 - b^2s^2 + s^2t^2 = 0$. Differentiating this with respect to s at constant t gives

$$4b^3\left(\frac{\partial b}{\partial s}\right)_t - 2bs^2\left(\frac{\partial b}{\partial s}\right)_t - 2b^2s + 2st^2 = 0$$

from which

$$\left(\frac{\partial b}{\partial s}\right)_t = \frac{s(b^2 - t^2)}{b(2b^2 - s^2)}$$

The other derivatives needed are derived in the solution to Prob. 1.1. Thus the r.h.s. is

$$\frac{bs}{(s^2 - b^2)^{1/2}} + \frac{s^2 - 2b^2}{(s^2 - b^2)^{1/2}} \cdot \frac{s(b^2 - t^2)}{b(2b^2 - s^2)} = \frac{bs}{h} - \frac{\frac{s}{b}(b^2 - t^2)}{h} = \frac{st^2}{bh} = t$$

where the last step involves $bh = st$. Thus the two sides agree.

1.3 We need to show that $(\partial A/\partial s)_b(\partial s/\partial b)_A(\partial b/\partial A)_s = -1$

From Eq. (1.2) $A = b(s^2 - b^2)^{1/2}$ we get by direct differentiation

$$\left(\frac{\partial A}{\partial s}\right)_b = \frac{bs}{(s^2 - b^2)^{1/2}}$$

For the others we use implicit differentiation on Eq. (1.2):

$$0 = (s^2 - b^2)^{1/2} - \frac{b^2}{(s^2 - b^2)^{1/2}} + \frac{bs}{(s^2 - b^2)^{1/2}}\left(\frac{\partial s}{\partial b}\right)_A$$

whence $(\partial s/\partial b)_A = (2b^2 - s^2)/bs$, and

$$1 = \left(\frac{\partial b}{\partial A}\right)_s (s^2 - b^2)^{1/2} - \frac{b^2}{(s^2 - b^2)^{1/2}}\left(\frac{\partial b}{\partial A}\right)_s$$

from which we find $(\partial b/\partial A)_s = (s^2 - b^2)^{1/2}/(s^2 - 2b^2)$. Putting this all into the equation above gives

$$\left(\frac{\partial A}{\partial s}\right)_b\left(\frac{\partial s}{\partial b}\right)_A\left(\frac{\partial b}{\partial A}\right)_s = \left[\frac{bs}{(s^2 - b^2)^{1/2}}\right]\left[\frac{2b^2 - s^2}{bs}\right]\left[\frac{(s^2 - b^2)^{1/2}}{s^2 - 2b^2}\right] = -1$$

1.4. In Eq. (1.1) we have $A(b,h) = bh$. Thus $A(xb,xh) = (xb)(xh) = x^2bh = x^2A(b,h)$. Also $b(\partial A/\partial b)_h + h(\partial A/\partial h)_b = bh + hb = 2A(b,h)$.

From Eq. (1.2) $A(b,s) = b(s^2 - b^2)^{1/2}$; therefore $A(xb,xs) = (xb)[(xb)^2 - (xs)^2]^{1/2} = xb[x^2(s^2 - b^2)]^{1/2} = x^2b(s^2 - b^2)^{1/2} = x^2A(s,b)$. Taking the derivatives from the solution to Prob. 1.3, we get

$$b\left(\frac{\partial A}{\partial b}\right)_s + s\left(\frac{\partial A}{\partial s}\right)_b = \frac{b(s^2 - 2b^2) + s(bs)}{(s^2 - b^2)^{1/2}} = \frac{2bs^2 - 2b^3}{(S^2 - b^2)^{1/2}} = 2b(s^2 - b^2)^{1/2}$$

$$= 2A(b,s)$$

From Eq. (1.3) $A(b,t) = b^2t/(b^2 - t^2)^{1/2}$, and $A(xb,xt) = (xb)^2(xt)/[(xb)^2 - (xt)^2]^{1/2} = x^3b^2t/[x^2(b^2 - t^2)]^{1/2} = x^2A(b,t)$.

$$b\left(\frac{\partial A}{\partial b}\right)_t + t\left(\frac{\partial A}{\partial t}\right)_b = b\left[\frac{2bt}{(b^2 - t^2)^{1/2}} - \frac{b^3t}{(b^2 - t^2)^{3/2}}\right]$$

$$+ t\left[\frac{b^2}{(b^2 - t^2)^{1/2}} + \frac{b^2t^2}{(b^2 - t^2)^{3/2}}\right] = \frac{3b^4t - 3b^2t^3 + b^2t^3 - b^4t}{(b^2 - t^2)^{3/2}} =$$

$$= \frac{2b^2t(b^2 - t^2)}{(b^2 - t^2)^{3/2}} = \frac{2b^2t}{(b^2 - t^2)^{1/2}} = 2A(b,t)$$

Finally, (1.4) can be treated like (1.1).

1.5 From $2xy\,dx + x^2\,dy$ we get

$$\frac{\partial}{\partial y}(2xy) = 2x = \frac{\partial}{\partial x}(x^2)$$

Therefore this is an exact differential. To find its integral from (0,0) to (1,1) along three paths we find:

Path 1, st. line $y=x$: Substituting x for y (and dx for dy) gives

$$\int_0^1 3x^2\,dx = x^3\Big]_0^1 = 1$$

Path 2, parabola $y=x^2$:

$$\int_0^1 [2x^3\,dx + x^2(2x\,dx)] = 4\int_0^1 x^3\,dx = 1$$

Path 3, $y=0$ followed by $x=1$: Along the first part $y = dy = 0$, and so the integral vanishes. Along the second part $dx = 0$, and so only the second term in the integrand remains. This gives

$$\int_0^1 dy = 1$$

Thus the integral is the same along all these paths. (The integrand is the differential of x^2y.)

1.6 Using the same three paths, find the integral of $2xy\,dx + x^2(y - 1)\,dy$ from (0,0) to (1,1).

$$\frac{\partial}{\partial y}(2xy) = 2x \neq 2x(y - 1) = \frac{\partial}{\partial x}[x^2(y - 1)]$$

Therefore this is not an exact differential, and its integral along the three paths yields the following values:

Path 1, $y=x$:

$$\int_0^1 (2x^2 + x^3 - x^2)\,dx = \left[\frac{x^3}{3} + \frac{x^4}{4}\right]_0^1 = \frac{7}{12}$$

Path 2, $y = x^2$:

$$\int_0^1 [2x(x^2)\,dx + x^2(x^2 - 1)(2x\,dx)] = \int_0^1 2x^5\,dx = \frac{1}{3}$$

Path 3, $y=0$ followed by $x=1$:

$$\int_0^1 (y - 1)\,dy = \frac{1}{2} - 1 = -\frac{1}{2}$$

This illustrates that for an expression that is not an exact differential the integral depends on the path.

2.1 a. $\Delta U = Q + W$ and $W = W' - p\Delta V$, which equals W' at constant V. Therefore $Q = \Delta U - W'$.

b. At constant p, $W - W' = -\int p\,dV = -p\Delta V = -\Delta(pV)$, and so $Q = \Delta U - W = \Delta U + \Delta(pV) - W' = \Delta H - W'$.

c. When $Q = 0$, $\Delta U = W = -\int p\,dV + W'$, and so the pressure-volume work $-\int p\,dV = \Delta U - W'$.

2.2 At constant pressure $dQ = dH$; in a reversible change $\Delta S = \int dQ/T = \int dH/T$. When T is constant this can be integrated to $\Delta S = (1/T)\int dH = \Delta H/T$.

2.3 At constant volume, $pdV = 0$ and so $dQ = dU$. Therefore

$$C_v = \left(\frac{\partial Q}{\partial T}\right)_V = \left(\frac{\partial U}{\partial T}\right)_V$$

Similarly, when p is constant, $dQ = dH$ and $C_p = (\partial H/\partial T)_p$. Generally, $dQ = T\,dS$ and so $C_X = (dQ/dT)_X = T(\partial S/\partial T)_X$. When X is S, this becomes $C_S = T(\partial S/\partial T)_S = 0$. Thus in isentropic changes (such as adiabatic reversible changes), the temperature can change without the addition or removal of heat.

2.4. $dQ = C_V\,dT$ (at constant volume) or $C_p\,dT$ (at constant pressure. Therefore $dS = dq/T = (C_V/T)\,dT$ or $(C_p/T)\,dT$ respectively.

2.5 By the definition of C_p and C_V

$$C_p - C_v = \left(\frac{\partial H}{\partial T}\right)_p - \left(\frac{\partial U}{\partial T}\right)_V = \left(\frac{\partial H}{\partial T}\right)_p - \left[\frac{\partial(H - pV)}{\partial T}\right]_V = \left(\frac{\partial H}{\partial T}\right)_p - \left(\frac{\partial H}{\partial T}\right)_V + V\left(\frac{\partial p}{\partial T}\right)_V$$

But from Eq. (1.7)

$$\left(\frac{\partial H}{\partial T}\right)_V = \left(\frac{\partial H}{\partial T}\right)_p + \left(\frac{\partial H}{\partial p}\right)_T \left(\frac{\partial p}{\partial T}\right)_V$$

Substituting this into the first equation gives

$$C_p - C_V = -\left(\frac{\partial H}{\partial p}\right)_T \left(\frac{\partial p}{\partial T}\right)_V + V\left(\frac{\partial p}{\partial T}\right)_V$$

which is equivalent to the required form. In a similar manner

CHAPTER 2

$$c_p - c_V = \left[\frac{\partial(U + pV)}{\partial T}\right]_p - \left(\frac{\partial U}{\partial T}\right)_V = \left(\frac{\partial U}{\partial T}\right)_p - \left(\frac{\partial U}{\partial T}\right)_V + p\left(\frac{\partial V}{\partial T}\right)_p$$

$$= \left(\frac{\partial U}{\partial V}\right)_T\left(\frac{\partial V}{\partial T}\right)_p + p\left(\frac{\partial V}{\partial T}\right)_p$$

2.6 a.

$$\frac{\kappa_T}{\kappa_S} = \frac{\left(\frac{\partial V}{\partial p}\right)_T}{\left(\frac{\partial V}{\partial p}\right)_S} = \frac{\left(\frac{\partial V}{\partial T}\right)_p}{\left(\frac{\partial p}{\partial T}\right)_V} \cdot \frac{\left(\frac{\partial S}{\partial V}\right)_p}{\left(\frac{\partial S}{\partial p}\right)_V}$$ [by applying Eq. 1.9 to both derivatives]

By a formula from elementary calculus this transforms into

$$\frac{(\partial S/\partial T)_p}{(\partial S/\partial T)_V} = \frac{c_p}{c_V}$$

2.6 b. From Eq. (1.7) we find

$$\kappa_S - \kappa_T = -\frac{1}{V}\left[\left(\frac{\partial V}{\partial p}\right)_S - \left(\frac{\partial V}{\partial p}\right)_T\right] = \frac{1}{V}\left(\frac{\partial V}{\partial S}\right)_p\left(\frac{\partial S}{\partial p}\right)_T$$

2.7 a. Of the heat supplied to the boiler 32% appears as electrical energy, and all of this goes to heat the buliding.

b. Because of loss up the chimney, only 65% of the heat is available in the building.

c. The electrical energy constitutes 32% of the heat, and this is available for work in the heat pump; with a coefficient of performance of 2.9, $|Q/W| = 2.9$, $W = 32\%$, $Q = 92.8\%$.

2.8 a. Carnot efficiency= (550 - 400)/550 = 27.3%

Actual efficiency = 0.75 x 27.3% = 20.45%

Useful heat = 100% - 20.45% = 79.55%

Power output = 2.5 MW = 20.45% of input heat.

Input heat = 2.5 MW/0.2045 = 12.22 MW

Output heat = 79.55% of 12.22 MW = 9.72 MW.

b. Efficiency = 0.75(550 - 310)/550 = 32.7%

Heat needed for power generation = 2.5 MW/0.327 = 7.64 MW

Heat needed for other use = 9.72 MW/0.65 = 14.95 MW

Total heat needed = 22.6 MW, contrasted with 12.2 MW with cogeneration.

CHAPTER 3

3.1 From $dU = T\,dS - p\,dV$ we get $(\partial T/\partial V)_S = -(\partial p/\partial S)_V$.
From $dH = T\,dS + V\,dp$ we get $(\partial T/\partial p)_S = (\partial V/\partial S)_p$.

3.2 a. By substituting $\Delta S = -(\partial \Delta A/\partial T)_V$ into $\Delta A = \Delta U - T\Delta S$ (at constant T) we get

$$\Delta A = \Delta U + T\left(\frac{\partial \Delta A}{\partial T}\right)_V$$

b. Similarly from $\Delta H = \Delta U + p\Delta V$ (at constant p) and $\Delta V = (\partial \Delta H/\partial p)_S$ we find

$$\Delta H = \Delta U + p\left(\frac{\partial \Delta H}{\partial p}\right)_S$$

c. From $\Delta U = \Delta A + S\Delta T$ (at constant S) and $\Delta T = (\partial \Delta U/\partial S)_V$ we get

$$\Delta U = \Delta A + S\left(\frac{\partial \Delta U}{\partial S}\right)_V$$

3.3 By Eq. (3.11) and the method in Sec. 1.3

$$\left(\frac{\partial S}{\partial V}\right)_T = \left(\frac{\partial p}{\partial T}\right)_V = -\frac{(\partial V/\partial T)_p}{(\partial V/\partial p)_T} = \frac{\alpha}{\kappa_T}$$

3.4. Using Eq. (1.9) (with derivatives inverted) and the result of Prob. 3.3, we find

$$\left(\frac{\partial T}{\partial V}\right)_S = -\frac{(\partial S/\partial V)_T}{(\partial S/\partial T)_V} = -\frac{\alpha T}{\kappa_T C_v}$$

Since T, κ_T, and C_v are all positive, the sign of $(\partial T/\partial V)_S$ is opposite that of α.

3.5 a. $dU = T\,dS - p\,dV$; divide by dV at constant T to get

$$\left(\frac{\partial U}{\partial V}\right)_T = T\left(\frac{\partial S}{\partial V}\right)_T - p = T\left(\frac{\partial p}{\partial T}\right)_V - p$$

where Eq. (3.11) has been used in the last step.

b. Similarly, we divide $dA = -S\,dT - p\,dV$ by dV and apply the result of Prob. 3.1 to get

$$\left(\frac{\partial A}{\partial V}\right)_S = -S\left(\frac{\partial T}{\partial V}\right)_S - p = S\left(\frac{\partial p}{\partial S}\right)_V - p$$

CHAPTER 3

3.6 Eq. (2.13) is

$$C_p - C_v = \left(\frac{\partial V}{\partial T}\right)_p \left[\left(\frac{\partial U}{\partial V}\right)_T + p\right]$$

but $\left(\frac{\partial V}{\partial T}\right)_p = V\alpha$ and by Prob. 3.5 $\left(\frac{\partial U}{\partial V}\right)_T + p = T\left(\frac{\partial p}{\partial T}\right)_V$

By Prob. 3.3 this last derivative is equal to α/κ_T; using this yields

$$C_p - C_v = \frac{TV\alpha^2}{\kappa_T}$$

In Prob. 2.6b we showed that

$$\kappa_T - \kappa_S = -\frac{1}{V}\left(\frac{\partial V}{\partial S}\right)_p \left(\frac{\partial S}{\partial p}\right)_T$$

By Eq. (3.10) the last derivative is equal to $-V\alpha$, and

$$\left(\frac{\partial V}{\partial S}\right)_p = \frac{(\partial V/\partial T)_p}{(\partial S/\partial T)_p} = \frac{V\alpha}{C_p/T} = \frac{VT\alpha}{C_p}$$

Putting all this together, we find $\kappa_T - \kappa_S = TV\alpha^2/C_p$.

3.7 $dU = T\,dS + \tau\,d\underline{l}$; $dH = T\,dS - \underline{l}\,d\tau$; $dA = -S\,dT + \tau\,d\underline{l}$; and $dG = -S\,dT - \underline{l}\,d\tau$. The Maxwell-type equations are

$$\left(\frac{\partial T}{\partial \underline{l}}\right)_S = \left(\frac{\partial \tau}{\partial S}\right)_{\underline{l}}; \quad \left(\frac{\partial T}{\partial \tau}\right)_S = -\left(\frac{\partial \underline{l}}{\partial S}\right)_\tau; \quad \left(\frac{\partial S}{\partial \underline{l}}\right)_T = -\left(\frac{\partial \tau}{\partial T}\right)_{\underline{l}}; \quad \text{and} \quad \left(\frac{\partial S}{\partial \tau}\right)_T = \left(\frac{\partial \underline{l}}{\partial T}\right)_\tau$$

The next two parts are so closely analogous to the treatments in Sec. 3.5 and 3.6 that they need not be given in detail. Dividing the rubber band into two identical parts, we find that

$$\delta S = \left(\frac{\partial^2 S}{\partial U^2}\right)_{\underline{l}} \delta U^2 + \ldots < 0$$

since an unnatural change is involved. Thus the second derivative must be negative, and

$$\left(\frac{\partial^2 S}{\partial U^2}\right)_{\underline{l}} = \left[\frac{\partial}{\partial U}\left(\frac{\partial S}{\partial U}\right)_{\underline{l}}\right]_{\underline{l}} = \left(\frac{\partial}{\partial U}\frac{1}{T}\right)_{\underline{l}} = -\frac{1}{T^2}\left(\frac{\partial T}{\partial U}\right)_{\underline{l}} < 0$$

whence $(\partial T/\partial U)_{\underline{l}}$ and so also its reciprocal $(\partial U/\partial T)_{\underline{l}}$ are positive.

Similarly, as in Sec. 3-6,

$$\delta A = \left(\frac{\partial^2 A}{\partial \underline{l}^2}\right)_T \delta \underline{l}^2 + \ldots > 0$$

and so

$$\left(\frac{\partial^2 A}{\partial \underline{l}^2}\right)_T = \left[\frac{\partial}{\partial \underline{l}}\left(\frac{\partial A}{\partial \underline{l}}\right)_T\right]_T = \left(\frac{\partial \tau}{\partial \underline{l}}\right)_T > 0$$

By analogy with Eqs. (1.8) or (1.9) we can write

$$\left(\frac{\partial T}{\partial \underline{l}}\right)_S = -\frac{\left(\frac{\partial S}{\partial \underline{l}}\right)_T}{\left(\frac{\partial S}{\partial T}\right)_{\underline{l}}} = \frac{\left(\frac{\partial \tau}{\partial T}\right)_{\underline{l}}}{C_{\underline{l}}/T} \text{ (by the 3rd Maxwell-type eq.) } = -\frac{T\left(\frac{\partial \underline{l}}{\partial T}\right)_\tau}{C_{\underline{l}}\left(\frac{\partial \underline{l}}{\partial \tau}\right)_T}$$

The statement that rubber shortens when heated at constant tension means that $(\partial \underline{l}/\partial T)_\tau < 0$; since $C_{\underline{l}}$ and $(\partial \underline{l}/\partial \tau)_T$ have been shown to be positive, we find that $(\partial T/\partial \underline{l})_S > 0$, as required.

3.8. In Prob. 3.1 we find that $(\partial T/\partial V)_S$ has dimensions $\theta/\underline{l}^3$, while the dimensions of $(\partial p/\partial S)_V$ are those of p $(m\underline{l}^{-1}t^{-2})$ divided by those of S $(m\underline{l}^2t^{-2}\theta^{-1})$. On carrying out the division the powers of m and t cancel out, and we are left with $\theta/\underline{l}^3$, so the two sides agree.

In 3.2a the short-cut dimensions of U and A are E, while those of the derivative $(\partial \Delta A/\partial T)_V$ are E/θ; when this is multiplied by θ, it also comes out E.

Similarly in 3.2b the dimensions of the derivative are E/(dim. of p), and thus that of $p(\partial \Delta H/\partial p)_S$ is just E, as is that of the other terms. 3.2c is treated similarly.

In 3.3 the dimensions of $(\partial S/\partial V)_T$ are $(E/\theta)/V = p/\theta$; those of α and κ_T are $1/\theta$ and $1/p$ respectively, and so the ratio is p/θ also.

In 3.4 the dimensions of $T\alpha$ are those of $(T/V)(\partial V/\partial T)_p$; these cancel out, so this combination is dimensionless. The product $C_V\kappa_T$ yields $(E/\theta)(1/p) = V/\theta$, and so the r.h.s. gives $1/(V/\theta) = \theta/V$, the same as the derivative $(\partial T/\partial V)_S$.

The term $T(\partial p/\partial T)_V$ in 3.5a has the same dimensions as p, as does the second term on the r.h.s. The derivative on the left gives $E/V = p$, and so the two sides agree. Similarly the term $S(\partial p/\partial S)_T$ in 3.5b, and the rest is the same as in 3.5a.

In 3.6a C_p and C_v are each of dimensions E/θ; on the r.h.s. α and κ_T have dimensions $1/\theta$ and $1/p$ respectively, and so the r.h.s. yields $\theta V(1/\theta)^2/(1/p) = pV/\theta = E/\theta$, in agreement with the l.h.s.

Using these same dimensions in 3.6b, we find $1/p$ for the l.h.s. and $\theta V(1/\theta)^2/(E/\theta) = V/E = 1/p$ on the right.

3.9 The fallacy lies in the assumption that an isentropic expansion can carry a system past a temperature at which the sign of its coefficient of expansion changes. The solution to Prob. 3.4 shows that when α reaches the value 0, the temperature does not change further.

4.1 $p = \alpha/V + \beta/V^2 + \gamma/V^3 + \ldots$; squaring and cubing this, retaining terms up to cubics, we find

$$p^2 = \frac{\alpha^2}{V^2} + \frac{2\alpha\beta}{V^3} + \ldots$$

$$p^3 = \frac{\alpha^3}{V^3} + \ldots$$

Substituting these into $pV = \alpha + Bp + Cp^2 + Dp^3 + \ldots$ gives

$$\left(\frac{\alpha}{V} + \frac{\beta}{V^2} + \frac{\gamma}{V^3} + \frac{\delta}{V^4} + \ldots\right)V = \alpha + B\left(\frac{\alpha}{V} + \frac{\beta}{V^2} + \frac{\gamma}{V^3} + \ldots\right) +$$

$$+ C\left(\frac{\alpha^2}{V^2} + \frac{2\alpha\beta}{V^3} + \ldots\right) + D\left(\frac{\alpha^3}{V^3} + \ldots\right)$$

The constant terms are α on each side. Equating the coefficients of $1/V$ gives $\beta = B\alpha = BRT$. The terms in V^{-2} yield $\gamma = B\beta + C\alpha^2$, and so, substituting the value of β, we find $\gamma = C\alpha^2 + B^2\alpha = C(RT)^2 + B^2RT$. Finally, from the cubic terms we get $\delta = B\gamma + 2C\alpha\beta + D\alpha^3 = B^3RT + 3BC(RT)^2 + D(RT)^3$.

4.2

$$g^\bullet(\text{atm}) = \lim_{p\to 0}\left(g - RT\ln\frac{p}{\text{atm}}\right)$$

$$g^\bullet(\text{bar}) = \lim_{p\to 0}\left(g - RT\ln\frac{p}{\text{bar}}\right)$$

On subtracting, we find that the difference on the r.h.s. is not dependent on p, so it is not necessary to take the limit. This gives

CHAPTER 4

$$g^\bullet(\text{atm}) - g^\bullet(\text{bar}) = RT \ln\frac{\text{atm}}{\text{bar}} = RT \ln 1.01325 = 32.6 \text{ J/mol}$$

4.3

$$p = \frac{RT}{v - b} - \frac{a}{v^2} = \frac{(0.083144 \frac{\text{L bar}}{\text{mol K}})(300 \text{ K})}{(2.00 - 0.04267) \text{ L/mol}} - \frac{3.545 \text{ L}^2\text{bar/mol}^2}{(2.00 \text{ L/mol})^2}$$

$$= 11.86 \text{ bars}$$

$$T = \frac{(p + \frac{a}{v^2})(v - b)}{R} = \frac{(4 \text{ bars} + \frac{3.545 \text{ L}^2\text{bar/mol}^2}{(20 \text{ L/mol})^2})(20 - 0.04267)\frac{\text{L}}{\text{mol}}}{0.083144 \text{ L bar/(mol K)}}$$

$$= 962.3 \text{ K}$$

4.4. Calculate the initial approximation by the ideal gas law:

$$v_0 = \frac{RT}{p} = \frac{(0.083144 \frac{\text{L bar}}{\text{mol K}})(310 \text{ K})}{25 \text{ bars}} = 1.0310 \text{ L}$$

The next approximation v_1 is found by substituting v_0 into the van der Waals equation:

$$v_1 = \frac{RT}{p + \frac{a}{v_0^2}} + b = 0.95230 \text{ L}$$

Successive approximations are then found by substituting each approximation into the vdW equation to get the next one. This leads to v_i/L (i = 2,3...) = (0.9342, 0.9296, 0.9283, 0.9280, 0.9279, 0.9279), so the final answer is 0.9279 L.

4.5 a. We start with the Berthelot equation in the form

$$p = \frac{RT}{v - b} - \frac{a}{Tv^2} \tag{1}$$

This is to be differentiated twice, and p_c, v_c, and T_c are to be substituted for p, v, and T. The derivatives are then set equal to zero:

$$\left(\frac{\partial p}{\partial v}\right)_{T(crit)} = -\frac{RT_c}{(v_c - b)^2} + \frac{2a}{T_c v_c^3} = 0 \qquad (2)$$

$$\left(\frac{\partial^2 p}{\partial v^2}\right)_{T(crit)} = \frac{2RT_c}{(v_c - b)^3} - \frac{6a}{T_c v_c^4} = 0 \qquad (3)$$

When the second term of each of these is transposed to the right, and then (2) is divided by (3), we find $(v_c - b)/2 = v_c/3$, whence $b = v_c/3$. From (2) [or (3)] we find

$$a = \frac{RT_c^2 v_c^3}{2(v_c - b)^2} = \frac{RT_c^2 v_c^3}{2\left(\frac{2}{3} v_c\right)^2} = \frac{9RT_c^2 v_c}{8} \qquad (4)$$

Then from (1) we get

$$p_c = \frac{RT_c}{\frac{2}{3} v_c} - \frac{9}{8}\frac{RT_c}{v_c} = \frac{3RT_c}{8v_c} \qquad (5)$$

from which $R = 8p_c v_c/3T_c$ and $a = 3p_c v_c^2 T_c$. Finally, substituting these into the Berthelot equation leads to

$$\left(p - \frac{3p_c v_c^2 T_c}{Tv^2}\right)\left(v - \frac{v_c}{3}\right) = \frac{8p_c v_c T}{3T_c}$$

which is easily reduced to

$$\left(p_r - \frac{3}{T_r v_r^2}\right)\left(3v_r - 1\right) = 8T_r$$

b. The equation may be written in the form

$$p = \frac{RT}{v - b} - \frac{a}{v^n} \qquad (6)$$

Setting the first and second derivatives, evaluated at the critical point, to zero gives

$$\frac{RT_c}{(v_c - b)^2} = \frac{na}{v_c^{n+1}} \qquad (7)$$

and

$$\frac{2RT_c}{(v_c - b)^3} = \frac{n(n + 1)a}{v_c^{n+2}} \tag{8}$$

Dividing (7) by (8) readily leads, after a bit of manipulation, to $b = (n-1)v_c/(n+1)$. Substitution into (7) or (8) then yields

$$a = \frac{RT_c v_c^{n+1}}{n(v_c - b)^2} = \frac{RT_c v_c^{n+1}}{n\left(\frac{2v_c}{n+1}\right)^2} = \frac{(n + 1)^2 RT_c v_c^{n-1}}{4n}$$

Substitution of this into the original equation gives

$$\left(p_c + \frac{(n + 1)^2 RT_c}{4nv_c}\right)\left(v_c - \frac{n - 1}{n + 1} v_c\right) = RT_c$$

and the solution for R is $4np_c v_c/[(n^2 - 1)T_c]$, which in turn leads to $a = (n + 1)p_c v_c^n/(n - 1)$. Inserted into the original equation, this leads after the usual manipulations to

$$\left[p_r + \frac{n + 1}{(n - 1)v_r^n}\right]\left[v_r - \frac{n - 1}{n + 1}\right] = \frac{4n}{n^2 - 1} T_r$$

c. From $p(v-b) = RT \exp(-a/RTv)$ we get

$$\ln p = \ln (RT) - \ln (v - b) - \frac{a}{RTv}$$

$$\left(\frac{\partial \ln p}{\partial v}\right)_T = \frac{1}{p} \left(\frac{\partial p}{\partial v}\right)_T$$

and

$$\left(\frac{\partial^2 \ln p}{\partial v^2}\right)_T = -\frac{1}{p^2} \left(\frac{\partial p}{\partial v}\right)_T + \frac{1}{p} \left(\frac{\partial^2 p}{\partial v^2}\right)_T$$

Since $1/p \neq 0$, it follows that the first and second derivatives of p are zero whenever those of ln p are; thus we can use the latter. Setting these derivatives, evaluated at the critical point, equal to zero, we find

$$\left(\frac{\partial \ln p}{\partial v}\right)_{crit} = -\frac{1}{v_c - b} + \frac{a}{RT_c v_c^2} = 0 \tag{9}$$

and

$$\left(\frac{\partial^2 \ln p}{\partial v^2}\right)_{crit} = \frac{1}{(v_c - b)^2} - \frac{2a}{RT_c v_c^3} = 0 \qquad (10)$$

As usual, division gives us the value of b, leading to $v_c - b = v_c/2$, whence $b = v_c/2$. Substitution of this into either (9) or (10) yields $a = 2RT_c v_c$. With these values the original Dieterici equation gives

$$p_c = \frac{RT_c}{v_c/2} e^{-2}, \text{ whence } R = \frac{e^2 p_c v_c}{2T_c} \text{ and } a = e^2 p_c v_c^2$$

Finally, the values of a, b, and R can be substituted into the Dieterici equation to get the reduced form:

$$p_r\,(2v_r - 1) = T_r \exp\left(2 - \frac{2}{v_r T_r}\right)$$

4.5d. It helps to write the Redlich-Kwong equation in the form

$$p = \frac{RT}{v - b} - \frac{a}{T^{1/2} b}\left[\frac{1}{v} - \frac{1}{v + b}\right]$$

Differentiating twice, evaluating the derivatives at the critical point, and setting them equal to zero leads to

$$\left(\frac{\partial p}{\partial v}\right)_{crit} = -\frac{RT_c}{(v_c - b)^2} + \frac{a}{T_c^{1/2} b}\left[\frac{1}{v_c^2} - \frac{1}{(v_c + b)^2}\right] = 0 \qquad (11)$$

$$\left(\frac{\partial^2 p}{\partial v^2}\right)_{crit} = \frac{2RT_c}{(v_c - b)^3} - \frac{a}{T^{1/2} b}\left[\frac{2}{v_c^3} - \frac{2}{(v_c + b)^3}\right] = 0 \qquad (12)$$

As usual, dividing one of these equations by the other eliminates R and a, giving us an equation for b:

$$\frac{v_c - b}{2} = \frac{\dfrac{1}{v_c^2} - \dfrac{1}{(v_c + b)^2}}{\dfrac{2}{v_c^3} - \dfrac{2}{(v_c + b)^3}}$$

From this we find

$$v_c - b = \frac{v_c(v_c + b)^3 - v_c^3(v_c + b)}{(v_c + b)^3 - v_c^3}$$

so that

$$b = v_c - \frac{v_c(v_c + b)^3 - v_c^3(v_c + b)}{(v_c + b)^3 - v_c^3} = \frac{v_c^3 b}{(v_c + b)^3 - v_c^3}$$

This reduces to

$$(v_c + b)^3 - v_c^3 = v_c^3$$

If we set $b = \rho v_c$, this becomes $(\rho + 1)^3 = 2$, so

$$\rho = 2^{1/3} - 1 = 0.25992...$$

By expanding the left side of the equation for ρ, we can derive the relation $\rho^3 + 3\rho^2 + 3\rho = 1$, which will be needed later. Setting $b=\rho v_c$ in (11) gives a value of a in terms of R and the critical constants:

$$\frac{RT_c}{v_c^2(1 - \rho)^2} = \frac{a}{\rho v_c T_c^{1/2}} \left[\frac{1}{v_c^2} - \frac{1}{(v_c + \rho v_c)^2}\right] = \frac{a}{\rho T_c^{1/2} v_c^3}\left[1 - \frac{1}{(1+\rho)^2}\right]$$

$$= \frac{a\rho(2 + \rho)}{\rho T_c^{1/2} v_c^3 (1 + \rho)^2}$$

from which we get

$$a = \frac{RT_c^{3/2} v_c (1 + \rho)^2}{(1 - \rho)^2 (2 + \rho)}$$

The denominator is $2 - 3\rho + \rho^3 = 2(\rho^3 + 3\rho^2 + 3\rho) - 3\rho + \rho^3$ [since the quantity in parentheses = 1] $= 3\rho^3 + 6\rho^2 + 3\rho = 3\rho(1 + \rho)^2$. Introducing this into the equation above gives $a = RT_c^{3/2} v_c / 3\rho = 1.28244 RT_c^{3/2} v_c$. We can then get the value of R by substituting the expressions for a and b into the Redlich-Kwong equation;

$$p_c = \frac{RT_c}{v_c(1 - \rho)} - \frac{RT_c^{3/2} v_c}{3\rho T_c^{1/2} v_c^2 (1 + \rho)} = \frac{RT_c}{v_c}\left(\frac{1}{1 - \rho} - \frac{1}{3\rho(1 + \rho)}\right)$$

$$= \frac{RT_c}{v_c}\left(\frac{3\rho^2 + 4\rho - 1}{3(\rho - \rho^3)}\right) = \frac{RT_c}{v_c}\left(\frac{3\rho^2 + 4\rho - 1}{3[\rho - (1 - 3\rho - 3\rho^2)]}\right) = \frac{RT_c}{3v_c}$$

With this we can eliminate R from the expression for a, obtaining

$$a = \left(\frac{3p_cv_c}{T_c}\right)\left(\frac{T_c^{3/2}v_c}{3\rho}\right) = \frac{p_cv_c^2T_c^{1/2}}{\rho} = 3.84732p_cv_c^2T_c^{1/2}$$

Substituting these into the Redlich-Kwong equation fimally permits us to convert it to reduced form:

$$p = \frac{3p_cv_cT}{T_c(v - \rho v_c)} - \frac{p_cT_c^{1/2}v_c^2}{\rho v(v + \rho v_c)T^{1/2}}$$

and after dividing both sides by p_c this is easily interpreted as

$$p_r = \frac{3T_r}{v_r - \rho} - \frac{1}{\rho v_r(v_r + \rho)T_r^{1/2}}$$

4.6 If the van der Waals equation is multiplied by v^2, the result is

$$(pv^2 + a)(v - b) = RTv^2$$

Expanding this and dividing by p gives (at the critical p & T)

$$v^3 - \frac{RT_c + p_cb}{p_c}v^2 + \frac{a}{p_c}v - \frac{ab}{p_c} = 0$$

Comparison with the equation $(v - v_c)^3 = v^3 - 3v_cv^2 + 3v_c^2v - v_c^3 = 0$

gives

$$\frac{RT_c + p_cb}{p_c} = 3v_c \qquad \frac{a}{p_c} = 3v_c^2 \quad \text{and} \quad \frac{ab}{p_c} = v_c^3$$

The second gives directly $a = 3p_cv_c^2$; this substituted into the third yields $b = v_c/3$, and finally the value of b in the first equation leads to $R = 8p_cv_c/(3T_c)$.

The treatment of the Berthelot equation is exactly the same, except that a is replaced throughout by a/T. This leads to the same results for b and R as the van der Waals equation, while a is given by $3p_cv_c^2T_c$.

Multiplying the Redlich-Kwong equation by $v(v - b)(v + b)$ gives

$$pv(v^2 - b^2) = RTv(v + b) - \frac{a}{T^{1/2}}(v - b)$$

which gives, after some algebraic manipulation and substitution of the critical values for p and T,

$$v^3 - \frac{RT_c}{p_c} v^2 + \frac{aT_c^{-1/2} - p_c b^2 - RT_c b}{p_c} v - \frac{ab}{p_c T_c^{1/2}} = 0$$

Comparison with $(v - v_c)^3 = 0$ then gives

$$\frac{RT_c}{p_c} = 3v_c \qquad \frac{a}{T_c^{1/2} p_c} - b^2 - \frac{RT_c}{p_c} b = 3v_c^2 \quad \text{and} \quad \frac{ab}{p_c T_c^{1/2}} = v_c^3$$

The first gives directly $R = 3p_c v_c/T_c$; according to the third the first term in the second is simply v_c^3/b, and putting this and the value of R into the second yields

$$\frac{v_c^3}{b} - b^2 - 3bv_c = 3v_c^2$$

With the substitution $b = \rho v_c$ this can be reduced to

$$\rho^3 + 3\rho^2 + 3\rho - 1 = 0$$

This can be solved numerically, but it is easier if we note that adding 2 to both sides converts the l. h. s. into $(\rho + 1)^3$. Therefore $\rho = 2^{1/3} - 1 = 0.25992$ as in Prob. 4.5. Then finally from the third equation $a = p_c T_c^{1/2} v_c^3/b = p_c T_c^{1/2} v_c^2/\rho$.

4.7. From $pV = \alpha + Bp + Cp^2 + \ldots$ we get $V = \alpha/p + B + Cp + \ldots$

$$\left(\frac{\partial V}{\partial p}\right)_T = -\frac{\alpha}{p^2} + C + \ldots$$

If the series converged at the critical point, this derivative would not be infinite, and so its reciprocal $(\partial p/\partial V)_T$, would not be zero as is required at the critical point.

4.8. a. The modified Berthelot equation can be solved for v:

$$v = \frac{RT}{p} + \frac{9RT_c}{128p_c}\left[1 - \frac{6}{T_r^2}\right] \qquad (1)$$

From this we get $(\partial v/\partial p)_T = -RT/p^2$, and so $(\partial p/\partial V)_T = -p^2/RT$, which is not zero at the critical point.

b. Putting the value of T_r into (1)

$$v = \frac{RT}{p} + \frac{9RT_c}{128p_c} - \frac{27RT_c^3}{64p_cT^2}$$

Differentiating gives

$$\left(\frac{\partial v}{\partial T}\right)_p = \frac{R}{p} + \frac{27RT_c^3}{32p_cT^3} = \frac{R}{p} + \frac{27R}{32p_cT_r^3}$$

Substituting this into Eq. (4.23), we find

$$s^\bullet = s_1 + \int_0^{p^\bullet} \frac{27R}{32p_cT_r^3}\,dp = s_1 + \frac{27Rp^\bullet}{32p_cT_r^3}$$

c. By integrating $(\partial s/\partial p)_T = -(\partial v/\partial T)_p$, noting that $s = s_1$ when $p = p^\bullet$, we get

$$s = s_1 - \int_{p^\bullet}^{p} \left(\frac{\partial v}{\partial T}\right)_p dp = s^\bullet - \frac{27Rp^\bullet}{32p_cT_r^3} - \int_{p^\bullet}^{p} \left(\frac{R}{p} + \frac{27R}{32p_cT_r^3}\right) dp$$

$$= s^\bullet - R\ln\frac{p}{p^\bullet} - \frac{27Rp}{32p_cT_r^3}$$

This gives a negative value for s at sufficiently high pressures, showing that the Berthelot equation can hold only at low pressures.

To evaluate h we integrate Eq. (4.4):

$$h - h^\bullet = \int_0^p \left[v - T\left(\frac{\partial v}{\partial T}\right)_p\right] dp = \int_0^p \left[\frac{9RT_c}{128p_c} - \frac{81RT_c^3}{64p_cT^2}\right] dp$$

the second form of the integrand being derived from expressions in part b. Since the integrand is independent of p, we need merely multiply by p; with a little manipulation this gives

$$h - h^\bullet = RT_c\left(\frac{9p_r}{128} - \frac{81p_r}{64T_r^2}\right)$$

Finally, referring again to part b for the integrand, we find

$$\ln\frac{f}{p} = \frac{1}{RT}\int_0^p \left(\frac{9RT_c}{128p_c} - \frac{27RT_c^3}{64p_cT^2}\right) dp = \frac{9p_r}{128T_r}\left(1 - \frac{6}{T_r^2}\right)$$

4.9. The approximations require setting $\Delta v = v_{gas} - v_{liq} \simeq v_{gas} \simeq RT/p$. Substituting this into the Clapeyron equation gives

$$\frac{dP}{dT} = \frac{\Delta h_{vap}}{T\Delta v} \simeq \frac{\Delta h_{vap}}{T(RT/P)} = \frac{P\Delta h_{vap}}{RT^2}$$

Trouton's rule gives the value of Δh_{vap} as (88 J mol^{-1} K^{-1})T, and P may be taken to be one bar (10^5 Pa). Substitution of these values gives

$$\frac{dP}{dT} = \frac{(10^5\ \text{Pa})(88\ \text{J mol}^{-1}\ \text{K}^{-1})T}{(8.3143\ \text{J mol}^{-1}\ \text{K}^{-1})T^2} \simeq \frac{1.06 \times 10^6\ \text{Pa}}{T} = \frac{10600\ \text{mbar}}{T}$$

and so

$$\frac{dT}{dP} \simeq \frac{T}{10600\ \text{mbar}}$$

That is, the boiling point (measured on the Kelvin scale) decreases by about one ten-thousandth of its value for each millibar decrease in pressure, if the pressure is in the neighborhood of standard atmospheric pressure.

4.10. With $p^\bullet$ = 1 bar = 100 kPa as the standard pressure, we make up the first four columns of the following table:

T/K	p/kPa	K/T	$\ln(p/p^\bullet)$	calc. p/kPa
200	8.649	0.005	-2.4477	8.681
210	17.749	0.0047619	-1.7288	17.656
220	33.814	0.0045454	-1.0843	33.664
230	60.442	0.0043478	-0.5035	60.685

By a linear regression program on the 3rd and 4th columns we get

$$\ln\frac{p}{p^\bullet} = -\frac{2981.6\ \text{K}}{T} + 12.464$$

The 5th column, calculated from this equation, shows good agreement (<0.6%) with the 2nd column, indicating that the equation is of satisfactory accuracy. Thus Δh_{vap} = (2981.6 K)R = 24790 J/mol. This should be rounded to 24.8 kJ/mol.

The same result (though with greater uncertainty) can be obtained graphically by plotting $\ln(p/p^\bullet)$ as ordinate against K/T as abscissa and measuring the slope to get the coefficient of K/T.

4.11. a. $pv = RT + Bp$. Plot pv vs. p, or use linear regression, to get RT = 23.53 L bars/mol, B = -0.15 L. This leads to R = 0.08311L bar/(mol K) = 8.311 J/(mol K).

b. It follows trivially from the equation given for h that $h^\bullet$ = lim h (as $p \to 0$) = 31.48 kJ/mol.

The expression for $s^\bullet$ is

$$s^\bullet = \lim_{p \to 0} (s + R \ln (p/p^\bullet))$$

The expression on the right has the values 211.67, 211.02, 210.21, 208.56, and 206.97 J/(K mol) at 1, 5, 10, 20, and 30 bars respectively. The limit can be found graphically or by linear regression; the latter yields s° = 211.83 J/(K mol).

Finally,

$$g^\bullet = h^\bullet - T s^\bullet = 31.48 \frac{\text{kJ}}{\text{mol}} - (283.15\ \text{K})(211.85 \frac{\text{J}}{\text{K mol}})(\frac{1\ \text{kJ}}{1000\ \text{J}})$$
$$= -28.50\ \text{kJ/mol}$$

c. At 30 bars h = (31.48 - 0.068x30) kJ/mol = 29.44 kJ/mol. The entropy is given as 178.69 J K^{-1} mol^{-1}. Therefore

$$g = 29.44\ \text{kJ/mol} - (283.15\ \text{k})(178.69\ \text{J K}^{-1}\ \text{mol}^{-1})(10^{-3}\ \text{kJ/J})$$
$$= -21.16\ \text{kJ/mol}$$

$$\frac{f}{p^\bullet} = \exp\left(\frac{g - g^\bullet}{RT}\right) = \exp\left(\frac{-21.16 - (-28.50)}{(8.3143)(283.15)} \times 10^3\right) = 22.6$$

$$\frac{f}{p} = \frac{f}{p^\bullet}\frac{p^\bullet}{p} = \frac{22.6}{30} = 0.75$$

The reduced variables are p_r = 0.406, T_r = 0.931; from the graph f/p is about 0.8.

d.

$$\mu_{JT} = \left(\frac{\partial T}{\partial p}\right)_h = -\frac{\left(\frac{\partial h}{\partial p}\right)_T}{\left(\frac{\partial h}{\partial T}\right)_p} = \frac{0.068\ \text{kJ mol}^{-1}\ \text{bar}^{-1}}{37.13\ \text{J K}^{-1}\ \text{mol}^{-1}} = 1.83 \frac{\text{K}}{\text{bar}}$$

e. Dividing the Berthelot equation by RT readily yields

$$z = 1 + \frac{9}{128}\frac{p_r}{T_r}\left(1 - \frac{6}{T_r^2}\right)$$

Substituting the values $p_r = 15/73.9 = 0.203$ and $T_r = 0.931$, we find $z = 0.909$. The value from the graph is 0.87

4.12. From dH = T dS + V dp and one of the Maxwell equations we get

$$\left(\frac{\partial h}{\partial v}\right)_T = T\left(\frac{\partial s}{\partial v}\right)_T + v\left(\frac{\partial p}{\partial v}\right)_T = T\left(\frac{\partial p}{\partial T}\right)_v + v\left(\frac{\partial p}{\partial v}\right)_T$$

Now

$$\left(\frac{\partial p}{\partial T}\right)_v = \frac{R}{v}\left(1 + \frac{B(T)}{v} + \frac{C(T)}{v^2} + \ldots\right) + \frac{RT}{v}\left(\frac{B'(T)}{v} + \frac{C'(T)}{v^2} + \ldots\right)$$

and

$$\left(\frac{\partial p}{\partial v}\right)_T = -\frac{RT}{v^2}\left(1 + \frac{B(T)}{v} + \frac{C(T)}{v^2} + \ldots\right) - \frac{RT}{v}\left(\frac{B(T)}{v^2} + \frac{2C(T)}{v^3} + \ldots\right)$$

Substitution gives

$$\left(\frac{\partial h}{\partial v}\right)_T = \frac{RT^2}{v}\left(\frac{B'(T)}{v} + \frac{C'(T)}{v^2} + \ldots\right) - RT\left(\frac{B(T)}{v^2} + \frac{2C(T)}{v^3} + \ldots\right)$$

$$= RT\left[\frac{TB'(T) - B(T)}{v^2} + \frac{TC'(T) - 2C(T)}{v^3} + \ldots\right]$$

Since $h \longrightarrow h^{\bullet}$ as $v \longrightarrow \infty$, we must use ∞ as the lower limit of integration. This leads to

$$h - h^{\bullet} = -RT\left[\frac{TB'(T) - B(T)}{v} + \frac{TC'(T) - 2C(T)}{2v^2} + \ldots\right]$$

The expressions given for B and C as functions of temperature give

$$B'(T) = -\frac{B_1}{T^2} - \frac{2B_2}{T^3} \quad \text{and} \quad C'(T) = -\frac{C_1}{T^2} - \frac{2C_2}{T^3}$$

Thus

$$h - h^{\bullet} = RT\left[\frac{1}{v}\left(B_0 + \frac{2B_1}{T} + \frac{3B_2}{T^2}\right) + \frac{1}{2v^2}\left(2C_0 + \frac{3C_1}{T} + \frac{4C_2}{T^2}\right)\right]$$

When the numerical values given by Holste et. al. are introduced, this reduces to

$$\frac{h - h^{\bullet}}{\text{J mol}^{-1}} = \left[\frac{\text{m}^3\ \text{mol}^{-1}}{v}\left(1.9148 \times 10^{-4}(T/K) - 4.2444 \times 10^{-2} - \frac{305.99}{T/K}\right)\right.$$
$$\left. + \frac{\text{m}^6\ \text{mol}^{-2}}{2v^2}\left(3.0620 \times 10^{-8}(T/K) - 2.5172 \times 10^{-5} + \frac{1.9063 \times 10^{-2}}{T/K}\right)\right]$$

Although the article gives values to 7 significant figures, they are rounded to 5 here to fit the precision to which R is known.

4.13 To use $z = pv/RT = 1 + B(T)/v + C(T)/v^2$ we need to calculate the value of B and C at 348.15 K:

$$B = \left(23.03 \times 10^{-6} - \frac{2.5525 \times 10^{-3}}{348.15} - \frac{12.2675}{348.15^2}\right)2 = -8.5512 \times 10^{-5}\ \text{m}^3/\text{mol}$$

In the same manner we find $C = 3.6718 \times 10^{-9}\ \text{m}^6\ \text{mol}^{-2}$. Then

$$z = 1 - \frac{8.5512 \times 10^{-5}}{1.1776 \times 10^{-3}} + \frac{3.6718 \times 10^{-9}}{(1.1776 \times 10^{-3})^2} = 0.9300$$

4.14. In Sec. 4.6 we found

$$dS = \frac{nR}{V}\,dV + \frac{C_v}{T}\,dT$$

We must eliminate either T or V. Choosing to retain T, we get

$$dV = \frac{V_2 - V_1}{T_2 - T_1}\,dT$$

and so

$$dS = \frac{nR}{V_1 + \frac{V_2 - V_1}{T_2 - T_1}(T - T_1)}\,\frac{V_2 - V_1}{T_2 - T_1}\,dT + \frac{C_v}{T}\,dT$$

$$= \left[\frac{nR(V_2 - V_1)}{V_1(T_2 - T_1) + (V_2 - V_1)(T - T_1)} + \frac{C_v}{T}\right]dT$$

We need to integrate this from T_1 to T_2; the result is

$$\Delta S = nR\ \ln\Big[V_1(T_2 - T_1) + (V_2 - V_1)(T - T_1)\Big]_{T_1}^{T_2} + C_v \ln\frac{T_2}{T_1}$$

When the limits T_1 and T_2 are substituted for T, a little manipulation reduces this to

$$\Delta S = nR \ln \frac{V_2}{V_1} + C_v \ln \frac{T_2}{T_1}$$

as derived in Sec. 4.6.

4.15. In the figure at the left we need to integrate v dp along the path BCDEF. Integrating by parts, we get

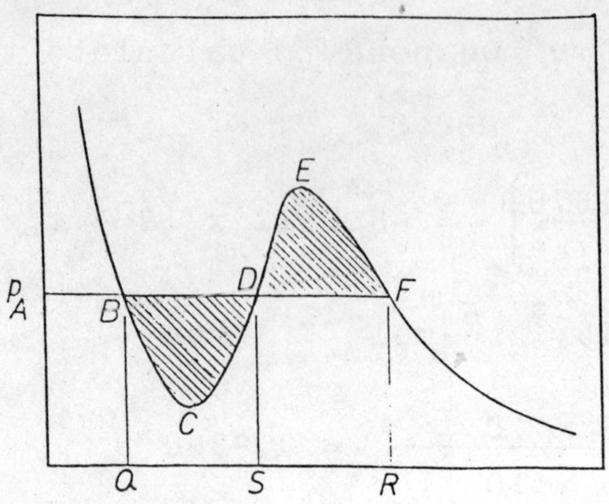

$$\Delta G = \int_B^F v\, dp = (pv)_F - (pv)_B - \int_{v_B}^{v_F} p\, dv$$

$= p_A(v_F - v_B)$ - area QBCDEFR

Separating $p_A(v_F - v_B)$ into the two rectangular areas $p_A(v_D - v_B)$ (= area QBDS) and $p_A(v_F - v_D)$ (= area SDFR) and area QBCDEFR into areas QBCDS and SDEFR, we find that QBDS - QBCDS is the shaded area BCD, while SDFR - SDEFR is the negative of the shaded area DEF. Thus for ΔG to be zero, these shaded areas must be equal.

The same conclusion can be reached by considering $\int v\, dp$ as representing the area to the left of the curve, extending to the p-axis, and assigning negative signs to areas adjacent to descending portions of the curve.

For a van der Waals gas

$$\int_{v_B}^{v_F} p\, dv = \int_{v_B}^{v_F} \left(\frac{RT}{v - b} - \frac{a}{v^2}\right) dv = RT \ln \frac{v_F - b}{v_B - b} + a\left(\frac{1}{v_F} - \frac{1}{v_B}\right)$$

When this is equated to $p_A(v_F - v_B)$, the required equation follows readily.

5.1. $Q = \Delta U = (100\ s)(10.00\ J/s) = 1000\ J$

$C_v = (1000\ J)/(2.012\ K) = 497.0\ J/K$

C_v (water) $= (50\ g)(4.184\ J\ K^{-1}\ g^{-1}) = 209.2\ J/K$

C (calorimeter) $= (497.0 - 209.2) J/K = 287.8\ J/K$

CHAPTER 5

5.2. We must convert the values given to values per uor by multiplying by the molar mass, 32.045 g/mol for N_2H_4 and 50.060 g/mol for $N_2H_4 \cdot H_2O$. This gives

	Δu/kJ uor^{-1}
N_2H_4 (l) + O_2 (g) $\longrightarrow$ N_2 (g) + $2H_2O$ (l)	-621.77
$N_2H_4 \cdot H_2O$ (l) + O_2 (g) $\longrightarrow$ N_2 (g) + $3H_2O$ (l)	-614.79

Subtracting the second from the first and rearranging yields

N_2H_4 (l) + H_2O (l) $\longrightarrow$ $N_2H_4 \cdot H_2O$ (l) -6.98

Δh is also -6.98 kJ/uor, since $\Delta n_g = 0$. For the first reaction

$$\Delta h = -621.77 \text{ kJ/uor} = 2\Delta h_f(H_2O;l) - \Delta h_f(N_2H_4;l)$$

and so

$$\Delta h_f(N_2H_4;l) = 2\Delta h_f(H_2O;l) + 621.77 \text{ kJ/uor}$$
$$= [2(-285.84) + 621.77] \text{ kJ/uor} = 50.09 \text{ kJ/mol}$$

Similarly from the second reaction we get

$$\Delta h_f(N_2H_4 \cdot H_2O;l) = 3\Delta h_f(H_2O;l) + 614.79 \text{ kJ/uor} = -242.73 \text{ kJ/mol}$$

5.3. At the transition temperature T_{tr} $\Delta g = 0$; at other temperatures $\Delta g = RT \ln [P(mo)/P(rh)]$, since the transition can be carried out reversibly by rhombic(s) $\longrightarrow$ gas [P(rh)] $\longrightarrow$ gas [P(mo)] $\longrightarrow$ mono.(s), and Δg for the 1st and 3rd steps is zero. By the Gibbs-Helmholtz equation

$$\left(\frac{\partial(\Delta g/T)}{\partial T}\right)_p = -\frac{\Delta h}{T^2}$$

and so

$$\frac{\Delta g}{T} = R \ln\left(\frac{\ln P(mo)}{\ln P(rh)}\right) = -\int_{T_{tr}}^{T} \frac{\Delta h}{T^2}\, dT$$

Since Δc_p is constant, $\Delta h(T)$ can be expressed in the form $\Delta h(T_0) + \Delta c_p(T - T_0)$; the figures given are for S instead of S_8 and so must be multiplied by 8. This gives

$$\Delta h(T) = 8\left(14.08 \text{ kJ/mol} + (23.6-22.6)\text{x}10^{-3}\text{kJ mol}^{-1}\text{ K}^{-1}(T - 368.65\text{ K})\right)$$
$$= \left(109.69 + 8 \text{ x } 10^{-3}(T/K)\right) \text{ kJ/mol}$$

Integration then leads to

$$R \ln \left(\frac{\ln P(mo)}{\ln P(rh)}\right) = \int_{298.15\ K}^{368.65\ K} \left(\frac{109.69\ kJ/mol}{T^2} + \frac{8 \times 10^{-3}\ kJ\ mol^{-1}\ K^{-1}}{T}\right) dT$$

$$= \left[- \frac{109.69\ kJ/mol}{T} \right]_{298.15\ K}^{368.65\ K} + 8 \times 10^{-3}\ kJ\ mol^{-1}\ k^{-1}\ \ln \frac{368.65}{298.15}$$

$$= 0.07205\ kJ\ mol^{-1}\ K^{-1} = 72.05\ J\ mol^{-1}\ K^{-1}$$

Finally, we find

$$\frac{P(mo)}{P(rh)} = \exp \left(\frac{72.05}{8.3143}\right) = 5.80 \times 10^3$$

5.4. For $CaO\ (s) + H_2O\ (l) \longrightarrow Ca(OH)_2\ (s)$

$$\Delta h = [-986.6 - (-636) - (-285.84)] kJ/uor = -64.8\ kJ/uor$$

$$\Delta g = [-896.8 - (-604) - (-237.14)]\ kJ/uor = -55.7\ kJ/uor$$

$$\Delta s = [76 - 40 - 69.9]\ J\ K^{-1}\ uor^{-1} = -33.9\ J\ K^{-1}\ uor^{-1}$$

$$\Delta h - T\ \Delta s = -64.8\ kJ/uor - (298.15\ K)(-33.9\ J\ K^{-1}\ uor^{-1})(10^{-3} kJ/J)$$

$$= -54.7\ kJ/uor\ (vs.\ -55.7)$$

For the reaction $2HI\ (g) + Cl_2\ (g) \longrightarrow 2HCl\ (g) + I_2\ (s)$

$$\Delta h = 2[-92.3 - 25.9] kJ/uor = -236.4\ kJ/uor$$

$$\Delta g = 2[-95.3 - 1.3]\ kJ/uor = -193.2\ kJ/uor$$

$$\Delta s = [117 + 2(186.78) - 222.95 - 2(206.44)]\ J\ K^{-1}\ uor^{-1}$$

$$= -145.27\ J\ K^{-1}\ uor^{-1}$$

$$\Delta h - T\ \Delta s = -236.4\ kJ/uor - (298.15\ K)(-145.49\ J\ K^{-1}\ uor^{-1})(10^{-3}\ kJ/J)$$

$$= -193.09\ kJ/uor\ (vs.\ -193.2)$$

5.5. $C_p\ \Delta T = (120\ J/K)(316.4\ K) = 37968\ J$

Electrical energy	900
Net energy from reaction	37068 J

Since the oxygen is present in excess, the amount of reaction is determined by the magnesium and is 1.50 g/24.05 g/mol = 0.06237 mol. Therefore Δu_f is (-37068 J/0.06237 mol)(10^{-3} kJ/J) = -594.3 kJ/mol.

$\Delta n_g = -1/2$ mol/uor, so

$$\Delta h_f = [-594.3 + (-1/2)(298.15)(8.3143)(10^{-3})]\ kJ/mol = -595.6\ kJ/mol.$$

5.6. C_p for the mixture need not be calculated, since Δh for each gas can be calculated and combined. This gives

$$\Delta h_{C_3H_6} = \int_{273.15}^{443.0} \left(6.136 + 0.03068\ (T/K)\right)\, dT =$$

$$[6.136(443.0 - 273.15) + 0.01534(443.0^2 - 273.15^2)] = 2908.1 \text{ J/mol}$$

For N_2, C_p is constant, so merely a multiplication is needed:

$$\Delta h_{N_2} = 3.5(8.3143 \text{ J K}^{-1} \text{ mol}^{-1})(443.0 \text{ K} - 273.15 \text{ K}) = 4942.6 \text{ J/mol}$$

Since each unit of reaction involves one mole of cyclopropane and five moles of nitrogen, we have for Δh for the reaction

$$\Delta h = -[2908.1 + 5(4942.7)] \text{ J/uor} = 27621.3 \text{ J/uor or } 27.62 \text{ kJ/uor.}$$

5.7. This is similar to 5.6 except that one mole of ozone produces 1.5 moles of oxygen. Thus C_p for the products is $[(20)(2.5) + (1.5)(3.5)]R = 55.25\ R$ for each mole of ozone. Thus Δh for the reaction of ozone to oxygen is $-(55.25\ R)(309.1\ K) = -141990$ J per mole of ozone. Thus Δh_f for ozone is +142.0 kJ/mol.

5.8. The midpoint of the range is 531.5 °C. Since only constant c_p values are involved, it is permissible to use Celsius temperatures.

From the vapor pressure equation we get $\Delta h_{vap}(\underline{1};\ 531.5°C) = 12440\ R = 103430$ J/mol. We need not be concerned with the fact that the equilibrium pressure varies with temperature, since the vapor is to be treated as an ideal gas, whose enthalpy is independent of pressure. In the liquid sodium range we have

$$\Delta c_p = 1.5\ R - 30.79 \text{ J K}^{-1}\text{mol}^{-1} = -18.32 \text{ J K}^{-1}\text{mol}^{-1}$$

Therefore

$$\Delta h_{vap}(\underline{1};\ 98°C) = [103430 + (98 - 531.5)(-18.32)] \text{ J/mol} = 111372 \text{ J/mol}$$

Since $\Delta h(s \rightarrow g) = \Delta h(s \rightarrow \underline{1}) + \Delta h(\underline{1} \rightarrow g)$

$$\Delta h_{sub}(98°C) = (3050 + 111372) \text{ J/mol} = 114422 \text{ J/mol}$$

$$\Delta c_p(s \rightarrow g) = 1.5\ R - 29.08 \text{ J K}^{-1}\text{mol}^{-1} = -16.61 \text{ J K}^{-1}\text{mol}^{-1}$$

Thus $\Delta h_{sub}(25°C) = [114422 + (25 - 98)(-16.61)]$ J/mol = 115634 J/mol.

Since $\Delta n_g = 1$, $\Delta u_{sub}(25°C) = 115634 - (298.15\ K)R = 113155$ J/mol.

Finally, these should be rounded: $\Delta h = 116$ kJ/mol, $\Delta u = 113$ kJ/mol.

5.9. At 298.15 K we have

$$\frac{\Delta h}{\text{kJ mol}^{-1}} = -636 + (-393.5) - (-1206.9) = 177.4$$

$$\frac{\Delta g}{\text{kJ mol}^{-1}} = -604 + (-394.4) - (-1128.8) = 130.4$$

$$\frac{\Delta s}{\text{J K}^{-1}\ \text{mol}^{-1}} = 40 + 213.8 - 92.9 = 160.9$$

$$\frac{\Delta c_p}{\text{J K}^{-1}\ \text{mol}^{-1}} = 37.13 + 42.8 - 88.9 = -8.97$$

By either method we can use Δh as a function of T. Since Δc_p is to be taken as constant, we do not need an integration but only a multiplication:

$$\Delta h(T) = \Delta h(T_1) + (T - T_1)\Delta c_p = \Delta h(T_1) - T_1\Delta c_p + \Delta c_p T = A + BT$$

where

$$A = 177.4\ \text{kJ/mol} - (298.15\ \text{K})(-8.97 \times 10^{-3}\text{kJ K}^{-1}\ \text{mol}^{-1}) = 180.1\ \text{kJ/mol}$$

and

$$B = -8.97 \times 10^{-3}\ \text{kJ K}^{-1}\ \text{mol}^{-1}$$

Substituting 1000 K for T gives $\Delta h(1000\ K) = 171.1$ kJ/mol. Then

$$\Delta s(1000\ \text{K}) = \Delta s(298.15\ \text{K}) + \Delta c_p \ln \frac{1000}{298.15} = 150.0\ \text{J K}^{-1}\ \text{mol}^{-1}.$$

Finally, using $\Delta g = \Delta h - T\Delta s$, we get

$$\Delta g(1000\ \text{K}) = 171.1\ \text{kJ/mol} - (1000\ \text{K})(0.1500\ \text{kJ K}^{-1}\text{mol}^{-1}) = 21.1\ \text{kJ/mol}$$

Alternatively, we can integrate the Gibbs-Helmholtz equation:

$$\frac{\partial(\Delta g/T)}{\partial T} = -\frac{\Delta h}{T^2}$$

to get

$$\frac{\Delta g(T_2)}{T_2} - \frac{\Delta g(T_1)}{T_1} = -\int_{T_1}^{T_2} \frac{A + BT}{T^2}\, dT = A\left(\frac{1}{T_2} - \frac{1}{T_1}\right) - B \ln \frac{T_2}{T_1}$$

Substituting numerical values in this equation leads to $\Delta g(1000\ K) =$ 24.26 kJ/mol.

As will be shown in Chap. 8, a low Δg signifies a high equilibrium constant; thus the comparatively low Δg at 1000 K, as compared to 298.15 K, means a relatively high equilibrium constant at the upper temperature. That this is a consequence of a positive Δh is one aspect of LeChatelier's law.

5.10. Start with

$\left[\frac{\partial(\Delta g/T)}{\partial T}\right]_p = -\frac{\Delta h}{T^2}$ and integrate, using integration by parts on the right-hand side. This gives $\left[\frac{\Delta g/T}{T}\right]_{T_1}^{T_2} = \left.\frac{\Delta h}{T}\right]_{T_1}^{T_2} - \int_{T_1}^{T_2} \frac{1}{T}\left(\frac{\partial \Delta h}{\partial T}\right)_p dT$

and the required relation follows as soon as we recognize that $(\partial\Delta h/\partial T)_p = \Delta c_p$. Alternatively, we can start with

$\Delta s(T_2) - \Delta s(T_1) = \int_{T_2}^{T_1} \frac{\Delta c_p}{T} dT$ and substitute in $\frac{\Delta g}{T} = \frac{\Delta h}{T} - \Delta s$.

6.1. $H = \Sigma\gamma_i \ln \gamma_i = \Sigma[(N_i/N)\ln(N_i/N)] = (1/N)\Sigma(N_i \ln N_1 - N_i \ln N) = (1/N)H_N - \ln N$, since $\Sigma N_i = N$.

6.2

$$k\left(\frac{\partial(T \ln Z)}{\partial T}\right)_V = kT\left(\frac{\partial \ln Z}{\partial T}\right)_V + k \ln Z = \frac{\hat{U}}{T} + k \ln Z = \hat{S}$$

6.3. By Eq. (6.21) r_m is the smallest integer exceeding $(ny-1)/(1-y)$. Therefore

$$\frac{ny - 1}{1 - y} + 1 = \frac{(n - 1)y}{1 - y} \geq r_m > \frac{ny - 1}{1 - y}$$

and so $$\frac{(n - 1)yh\nu}{1 - y} \geq u_m > \frac{(ny - 1)h\nu}{1 - y} \qquad (1)$$

By Eq. (6.21) $$\bar{u} = \frac{nyh\nu}{1 - y} \qquad (2)$$

Subtracting (1) from (2) gives

$$\frac{nyh\nu}{1-y} - \frac{(n-1)yh\nu}{1-y} \leq \bar{u} - u_m < \frac{nyh\nu}{1-y} - \frac{(ny-1)h\nu}{1-y}$$

or

$$\frac{yh\nu}{1-y} \leq \bar{u} - u_m < \frac{h\nu}{1-y}$$

Dividing the middle member by $\bar{u}$ and the outer ones by its equivalent from (2) gives the required answer

$$\frac{1}{n} \leq \delta < \frac{1}{ny}$$

If $n = 1$, this gives $1 \leq \delta < 1/y$, but since $\delta \leq 1$, this means that $\delta = 1$. This reflects the fact that for a single oscillator the ground state is the most populated one; that is, $u_m = 0$, which necessarily leads to $\delta = 1$ unless $\bar{u}$ is also 0.

To find n for T = 100 K, ν = 2 thz, and $\delta \leq 10^{-6}$, note that

$$y = \exp\left(-\frac{h\nu}{kT}\right) = \exp\left[-\frac{(6.6262 \times 10^{-34}\,\mathrm{Js})(2 \times 10^{12}\,\mathrm{s}^{-1})}{(1.3806 \times 10^{-23}\,\mathrm{J/K})(100\ \mathrm{K})}\right] = 0.383$$

Now to make sure than δ does not exceed 10^{-6}, we must restrict $1/ny$ to this value; that is, $1/(0.383\ n) \leq 10^{-6}$, whence $n \geq 2.61 \times 10^6$. This amounts to about 4.3×10^{-18} mole.

6.4.

$$\hat{u} = \frac{3Nyh\nu}{1-y}$$

and

$$C_v = \frac{d\hat{u}}{dT} = \left[\frac{3Nh\nu}{1-y} + \frac{3Nyh\nu}{(1-y)^2}\right]\frac{dy}{dT} = \frac{3Nh\nu}{(1-y)^2}\left[y\frac{h\nu}{kT^2}\right] = \frac{3Nky}{(1-y)^2}\left(\frac{h\nu}{kT}\right)^2 \quad (1)$$

Now $y = \exp(-h\nu/kT)$, and as $T \to \infty$, $h\nu/kT \to 0$, and $y \to 1$. At the same time $1 - y$ behaves as

$$\left(\frac{h\nu}{kT}\right) - \mathcal{O}(T^{-2})$$

where $\mathcal{O}(T^{-2})$ means terms of quadratic or higher power in the reciprocal of T. For sufficiently high temperatures all terms except the first can be neglected; this makes the denominator in (1) $(h\nu/kT)^2$ and so $C_v = 3Nk =$ (for one mole) 3R.

As $T \to 0$, $h\nu/kT$ becomes very large, and $y = \exp(-h\nu/kT) \to 0$.

Thus we can neglect the y in the denominator of (1); then using x for $h\nu/kT$, we find $C_v = 3Nkx^2/e^x$. As $x \to \infty$, it is easily shown by the series expansion of e^x or by l'Hospital's rule that this expression approaches zero.

6.5. The easiest technique, though not the only one, is to use β instead of T. Noting that $d\beta/dT = -1/kT^2$, we find

$$C_v = \left(\frac{\partial \hat{u}}{\partial T}\right)_v = \left(\frac{\partial \hat{u}}{\partial \beta}\right) \frac{d\beta}{dT} = -\frac{1}{kT^2}\frac{\partial}{\partial \beta}\left(-\frac{\partial \ln Z}{\partial \beta}\right)_v = \frac{1}{kT^2}\frac{\partial}{\partial \beta}\left[\frac{\partial Z/\partial \beta}{Z}\right]$$

$$= \frac{1}{kT^2}\left[\frac{\partial^2 Z/\partial \beta^2}{Z} - \frac{(\partial Z/\partial \beta)^2}{Z^2}\right] = \frac{1}{kT^2}\left[\frac{1}{Z}\sum_i \varepsilon_i{}^2 e^{-\varepsilon_i/kT} - \left(\frac{\partial Z/\partial \beta}{Z}\right)^2\right]$$

$$= \frac{1}{kT^2}\left(\overline{u^2} - \overline{u}^2\right)$$

The quantity in parentheses in this last expression is the square of the standard deviation of the energy, and so the standard deviation itself is $(kT^2C_v)^{1/2}$. But the energy and C_v are extensive properties; that is, they are proportional to N; thus the relative standard deviation is proportional to $N^{-1/2}$.

6.6. Since the number in the lower state is $N - n_1$, we find

$$\hat{u} = (N - n_1)(-\varepsilon) + n_1\varepsilon = (2n_1 - N)\varepsilon$$

The number of ways that N particles can be distributed so that n_1 are in one state is the binomial coefficient $\binom{N}{n_1}$; thus the entropy is

$$\hat{s} = k \ln \binom{N}{n_1} \simeq k\{N \ln N - n_1 \ln n_1 - (N - n_1) \ln (N - n_1)\}$$

$$\frac{d\hat{u}}{dn_1} = 2\varepsilon \text{ and } \frac{d\hat{s}}{dn_1} = k\{-\ln n_1 - 1 + \ln (N - n_1) + 1\} = k \ln\left(\frac{N}{n_1} - 1\right)$$

and so
$$T = \frac{d\hat{u}}{d\hat{s}} = \frac{2\varepsilon}{k \ln\left(\frac{N}{n_1} - 1\right)}$$

Now when $n_1 = 0$, the denominator is infinite, and $T = 0$. When n_1 is slightly less than N/2, the argument of the logarithm is slightly more than one, and the logarithm is a small positive number; thus T is large and positive. At $n_1 = N/2$, the logarithm is zero, and T is undefined. At n_1 just above N/2, the logarithm is small and negative, and T is large and negative. Finally, as $n_1 \longrightarrow N$, the argument of the logarithm becomes a very small positive number; the logarithm is large and negative, and T approaches zero from below. Thus as n_1 goes from 0 to N, the temperature increases to ∞ at N/2, changes discontinuously to $-\infty$, and then increases to zero. Simultaneously β starts at ∞, decreases to zero at N/2, and goes to $-\infty$ at N, without any discontinuities. Negative temperatures cannot occur in systems with an infinite number of available states.

6.7.

$n_J = (2J + 1)e^{-J(J + 1)\sigma}/Z$ gives, on differentiation w. r. t. J,

$$\frac{dn_J}{dJ} = \left[2 - \sigma(2J + 1)^2\right]e^{-J(J + 1)\sigma}/Z$$

For a maximum we must have $2 - \sigma(2J + 1)^2 = 0$, which leads to

$$J = \frac{1}{(2\sigma)^{1/2}} - \frac{1}{2}$$

and since J must be an integer, we accept the integer nearest this as the most populated level.

6.8 For the transition from J = 0 to J = 1, the energy difference $\Delta\varepsilon$ is equal to h^2/I; from the data given we can calculate this as $(3.76 \times 10^{11}\ s^{-1})(6.6262 \times 10^{-34}\ Js) = 2.491 \times 10^{-22}\ J$. Thus

$$\frac{2Ik}{h^2} = \frac{2(1.3806 \times 10^{-23}\ J/K)}{2.491 \times 10^{-22}\ J} = 0.1108\ K^{-1}$$

At 25 °C the molar rotational entropy is $R\{1 + \ln[(0.1180\ K^{-1})(298.15\ K)]\} = 37.39\ J\ K^{-1}\ mol^{-1}$.

The vibrational energy is

$$\frac{N_0 h\nu}{e^{h\nu/kT} - 1} = \frac{(6.022 \times 10^{23}\ mol^{-1})(6.6262 \times 10^{-34} Js)(6.67 \times 10^{13}\ s^{-1})}{\exp\left[\frac{(6.6262 \times 10^{-34} Js)(6.67 \times 10^{13} s^{-1})}{(1.3806 \times 10^{-23} J/K)(298.15\ K)}\right] - 1}$$

$= 0.578$ J/mol

Thus the vibrational entropy is

$$\frac{0.578 \text{ J/mol}}{298.15 \text{ K}} - R \ln (1 - e^{-h\nu/kT}) = 2.12 \text{ mJ K}^{-1} \text{ mol}^{-1}$$

M for HI is 127.88 g/mol; using this we get

$$S^{\bullet}_{trans} = R\left[\frac{3}{2} \ln 127.88 + \frac{5}{2} \ln 298.15 - 1.1517\right] = 169.35 \text{ J K}^{-1} \text{ mol}^{-1}$$

Thus the vibrational entropy is negligible, and the total entropy is $(169.35 + 37.40)$ J K^{-1} mol $^{-1}$ = 206.75 J K^{-1} mol^{-1}.

At 1000 K the calculations are similar and lead to

$s_{rot} = 47.5$ J K^{-1}

$u_{vib} = 1129.5$ J mol^{-1}

$s_{vib} = 1.475$ J K^{-1} mol^{-1}

$s^{\bullet}_{trans} = 194.5$ J $K^{-1} mol^{-1}$.

$s^{\bullet}_{total} = 243.5$ J K^{-1} mol^{-1}.

6.9. Since

$$\omega(\nu) = \frac{9N}{\nu_m^3} \nu^2$$

the zero-point energy is

$$\int_0^{\nu_m} \frac{9N}{\nu_m^3} \nu^2 \left(\frac{1}{2}h\nu\right) d\nu = \frac{9Nh}{2\nu_m^3} \frac{\nu_m^4}{4} = \frac{9}{8} Nh\nu_m = \frac{9}{8}Nk\Theta$$

6.10. By Eq. (6.53) we have

$$\ln Z = -\int_0^{\nu_m} \omega(\nu) \ln (1 - y) \, d\nu$$

Make the substitutions $x = h\nu/kT$, $y = e^{-x}$, $\Theta = h\nu_m/k$ to get

$$\ln Z = -9N\left(\frac{T}{\Theta}\right)^3 \int_0^{\Theta/T} x^2 \ln (1 - e^{-x}) \, dx \qquad (1)$$

Now use integration by parts to transform the integral:

$$\int_0^{\Theta/T} x^2 \ln(1 - e^{-x})\, dx = \frac{x^3}{3} \ln(1 - e^{-x})\Big]_0^{\Theta/T} - \int_0^{\Theta/T} \frac{x^3}{3} \frac{e^{-x}}{1 - e^{-x}}\, dx$$

$$= \frac{1}{3}\left(\frac{\Theta}{T}\right)^3 \ln(1 - e^{-\Theta/T}) - \frac{1}{3}\int_0^{\Theta/T} \frac{x^3\, dx}{e^x - 1}$$

Substitution in (1) then gives

$$\ln Z = 3N\left(\frac{T}{\Theta}\right)^3 \int_0^{\Theta/T} \frac{x^3\, dx}{e^x - 1} - 3N \ln(1 - e^{-\Theta/T}) = \frac{U}{3kT} - 3N \ln(1 - e^{-\Theta/T})$$

by comparison with Eq. (6.58). Using the last expression in the equation $S = U/T + k \ln Z$ gives

$$S = \frac{4U}{3T} - 3Nk \ln(1 - e^{-\Theta/T})$$

In differentiating U with respect to T to get C_v, we need

$$\frac{d}{dT}\int_0^{\Theta/T} \frac{x^3\, dx}{e^x - 1} = \frac{(\Theta/T)^3}{e^{\Theta/T} - 1} \frac{d}{dT}\left(\frac{\Theta}{T}\right) = -\frac{\Theta^4}{T^5(e^{\Theta/T} - 1)}$$

Then we find

$$C_v = \frac{d}{dT}\left[\frac{9NkT^4}{\Theta^3}\int_0^{\Theta/T} \frac{x^3\, dx}{e^x - 1}\, dx\right] = \frac{36NkT^3}{\Theta^3}\int_0^{\Theta/T} \frac{x^3\, dx}{e^x - 1} - \frac{9NkT^4}{\Theta^3} \frac{\Theta^4}{T^5(e^{\Theta/T} - 1)}$$

$$= 36Nk\left(\frac{T}{\Theta}\right)^3 \int_0^{\Theta/T} \frac{x^3\, dx}{e^x - 1} - \frac{9Nk\Theta}{T(e^{\Theta/T} - 1)} \qquad (2)$$

Integration by parts gives

$$\int_0^{\Theta/T} \frac{x^3\, dx}{e^x - 1} = \frac{x^4}{4(e^x - 1)}\Big]_0^{\Theta/T} - \frac{1}{4}\int_0^{\Theta/T} x^4 \frac{d}{dx}\left(\frac{1}{e^x - 1}\right) dx$$

$$= \frac{(\Theta/T)^4}{4(e^{\Theta/T} - 1)} + \frac{1}{4}\int_0^{\Theta/T} \frac{x^4 e^x\, dx}{(e^x - 1)^2}$$

When this is substituted into (2), the integrated parts cancel out, and we find

$$C_V = 9Nk\left(\frac{T}{\Theta}\right)^3 \int_0^{\Theta/T} \frac{x^4 e^x \, dx}{(e^x - 1)^2}$$

6.11. Eq. (6.64), in slightly different form, is

$$U = \frac{8\pi^5 k^4 T^4}{15c^3 h^3} V$$

Differentiating with respect to T gives an expression for C_V:

$$C_V = \frac{32\pi^5 k^4 T^3}{15c^3 h^3} V$$

From this we find

$$S = \int_0^T \frac{C_V \, dT}{T} = \int_0^T \frac{32\pi^5 k^4 T^2}{15c^3 h^3} V \, dT = \frac{32\pi^5 k^4 T^3}{45c^3 h^3} V = \frac{4U}{3T}$$

$$A = U - TS = U - \frac{4}{3} U = -\frac{1}{3} U = -\frac{8\pi^5 k^4 T^4}{45c^3 h^3} V$$

$$p = -\left(\frac{\partial A}{\partial V}\right)_T = \frac{8\pi^5 k^4 T^4}{45c^3 h^3} = \frac{1}{3}\frac{U}{V}$$

Finally, we may note that

$$G = A + pV = -\frac{1}{3} U + \left(\frac{1}{3}\frac{U}{V}\right)V = 0$$

6.12. By the relativistic equivalence of mass and energy, a photon has mass $h\nu/c^2$ and so has momentum $h\nu/c$. If a photon is absorbed, its momentum is reduced to a negligible value($h\nu v/c^2$, where v is the speed of the paddle), and so momentum in the amount $h\nu/c$ is transferred to the paddle. However, if the photon is reflected, its momentum is reversed, resulting in the transference of momentum in the amount $2h\nu/c$ to the paddle. Thus radiation pressure is twice as great on the shiny side as on the dark side, and the direction of rotation must be with the dark side leading.

Actually, these devices rotate in the other direction, so radiation pressure cannot be the motive force.

The dark side absorbs more solar energy than the shiny one and so becomes warmer. A gas molecule striking the dark side therefore picks up extra energy, and rebounds with more force than one striking the shiny side. That this is the means of operation is shown not only by the direction of rotation, but by the fact that the devices will not function is the evacuation of the bulb is too thorough.

6.13. For the transition from J to J-1, we have

$$\Delta\varepsilon_{J\rightarrow J-1} = [J(J + 1) - (J - 1)J]\frac{h^2}{2I} = \frac{Jh^2}{4\pi^2 I}$$

and so

$$\nu_{J\rightarrow J-1} = \frac{Jh}{4\pi^2 I}$$

For the 2 lowest J values (J and J+1)

$$4.990 \text{ thz} = \frac{Jh}{4\pi^2 I}$$

and

$$5.103 \text{ thz} = \frac{(J + 1)h}{4\pi^2 I}$$

a. Dividing gives 1.0226 = (J + 1)/J, whence J = 44. Thus the initial J values for the transitions listed are 44, 45, ..., 49.

b.

$$\Delta\varepsilon_{44\rightarrow 43} = h\nu = (4.990 \times 10^{12}\ s^{-1})(6.6262 \times 10^{-34} Js) = 3.306 \times 10^{-21} J$$

The others, similarly calculated, are given by

$$\frac{\Delta\varepsilon}{10^{-21} J} = 3.381,\ 3.456,\ 3.532,\ 3.607,\ 3.682$$

c. For the J = 44 to J = 43 transition $3.306 \times 10^{-21} J = \frac{44h^2}{4\pi^2 I}$

d. This leads to $I = 1.480 \times 10^{-46}$ kg m^2

e. The reduced mass is given by $\frac{1}{\mu} = \frac{1}{m_C} + \frac{1}{m_N} = \frac{1}{12 \text{ amu}} + \frac{1}{14 \text{ amu}}$

whence μ = 6.462 amu = 1.073×10^{-26} kg. Now using $I = \mu r^2$ we get 1.480×10^{-46} kg m^2 = $(1.073 \times 10^{-26}$ kg$)r^2$, which gives $r = 1.174 \times 10^{-10}$ m = 0.1174 nm.

f.
$$\Delta\varepsilon_{44\to 49} = [(49)(50)-(44)(45)]\frac{\hbar^2}{2(1.480 \times 10^{-46}\text{kg m}^2)} = 1.766 \times 10^{-20} \text{ J}$$

Taking account of the degeneracy 2J + 1 of the energy levels, we find for the ratio of populations $\frac{2(49)+1}{2(44)+1}\exp\left(\frac{-1.776 \times 10^{-20}\text{J}}{kT}\right) = 0.65$

$$-\frac{1.776 \times 10^{-20}\text{ J}}{kT} = \ln(0.65) - \ln\frac{99}{89} = -0.5373$$

whence $T \simeq 2390$ K.

6.14. This follows closely the treatment of the 3-dimensional solid in the text. In 2 dimensions the wave equation is

$$\frac{\partial^2 u}{\partial x^2} + \frac{\partial^2 u}{\partial^2 y} = \frac{1}{c^2}\frac{\partial^2 u}{\partial t^2}$$

and setting $u(x,y,t) = X(x)Y(y)T(t)$ leads to

$$\frac{X''}{X} + \frac{Y''}{Y} = \frac{1}{c^2}\frac{T''}{T}$$

So we can set
$$X''/X = -\alpha^2; \quad Y''/Y = -\beta^2; \quad T''/(c^2T) = -\gamma^2$$

with $\alpha^2 + \beta^2 = \gamma^2$. If the dimensions of the sheet are a(in the x-direction) and b, the argument in Appendix 6 gives

$$\alpha a = \lambda\pi \quad \beta b = \mu\pi \text{ and } \gamma = 2\pi\nu/c$$

where λ and μ are integers. From this we find

$$\left(\frac{\lambda\pi}{a}\right)^2 + \left(\frac{\mu\pi}{b}\right)^2 = \left(\frac{2\pi\nu}{c}\right)^2$$

or

$$\frac{\lambda^2}{\left(\frac{2a\nu}{c}\right)^2} + \frac{\mu^2}{\left(\frac{2b\nu}{c}\right)^2} = 1$$

The number of modes of oscillation with frequency $\leq \nu$ is the number of points with integral coordinates in the positive quadrant of the ellipse represented by this equation; that is,

$$\Omega(\nu) = \frac{\pi}{4}\left(\frac{2a\nu}{c}\right)\left(\frac{2b\nu}{c}\right) = \frac{\pi A}{c^2}\nu^2$$

where A = ab is the area of the sheet.
Since the number of oscillators is N, $\Omega(\nu_m) = N = \pi A\nu_m^2/c^2$. Substituting this into the equation above gives

$$\Omega(\nu) = \frac{N}{\nu_m^2}\nu^2, \text{ and } \omega(\nu) = \Omega'(\nu) = \frac{2N\nu}{\nu_m^2}$$

$$\ln Z = \int_0^{\nu_m} \omega(\nu) \ln (1 - y)\, d\nu = \frac{2N}{\nu_m^2}\int_0^{\nu_m} \nu \ln (1 - y)\, d\nu$$

From this point the treatment is so closely analogous to that in Sec. 6.14 that details can be skipped. Following that treatment as a model, including the introduction of the characteristic temperature $\Theta = h\nu_m/k$, we find

$$U = \frac{2NkT^3}{\Theta^2}\int_0^{\Theta/T} \frac{x^2dx}{e^x - 1} \simeq \text{(for low T)}\ \frac{2NkT^3}{\Theta^2}\int_0^{\infty} \frac{x^2dx}{e^x - 1}$$

Using the value 2.40 given for the last integral, we get by differentiating w. r. t. T,

$$C_v = \frac{6NkT^2}{\Theta^2}(2.40) = 14.4Nk\left(\frac{T}{\Theta}\right)^2$$

6.15. The entropy of mixing of α mole of ^{35}Cl and β mole of ^{37}Cl (with $\alpha + \beta = 1$) would be $\Delta S_I = -R(\alpha \ln \alpha + \beta \ln \beta)$. Dimerization of two moles of Cl would produce α^2 mole of $(^{35}Cl)_2$, $2\alpha\beta$ mole of $(^{35}Cl)(^{37}Cl)$, and β^2 mole of $(^{37}Cl)_2$. Note the coefficient 2 for the heteronuclear molecule; this results from the fact that either of the two atoms may be the ^{37}Cl. The necessity of this factor can also be seen by the fact that the amounts of the three species must total 1, and $\alpha^2 + 2\alpha\beta + \beta^2 = (\alpha + \beta)^2 = 1$. Since only half this much atomic chlorine is involved, the amounts of the molecular forms are half these. Therefore, the entropy of mixing of the molecules is

$$\begin{aligned}\Delta S_{II} &= -\frac{R}{2}[\alpha^2 \ln (\alpha^2) + 2\alpha\beta \ln (2\alpha\beta) + \beta^2 \ln (\beta^2)]\\ &= -R[(\alpha^2 + \alpha\beta) \ln \alpha + (\alpha\beta + \beta^2) \ln \beta + \alpha\beta \ln 2]\\ &= -R(\alpha \ln \alpha + \beta \ln \beta + \alpha\beta \ln 2) \quad [\text{Since } \alpha^2+\alpha\beta=\alpha(\alpha+\beta)=\alpha, \text{ etc.}]\\ &= \Delta S_I - R\alpha\beta \ln 2\end{aligned}$$

But $\alpha\beta R \ln 2$ is precisely the extra amount of rotational entropy that the $^{35}Cl^{37}Cl$ molecules have because of their heteronuclear character. Adding this amount to ΔS_{II} brings it into agreement with ΔS_I.

6.16. The possible spin states are.

Sym	Antisym
$\alpha\alpha$	$\alpha\beta - \beta\alpha$
$\beta\beta$	$\alpha\gamma - \gamma\alpha$
$\gamma\gamma$	$\beta\gamma - \gamma\beta$
$\alpha\beta + \beta\alpha$	
$\alpha\gamma + \gamma\alpha$	
$\beta\gamma + \gamma\beta$	

a. Since there are six symmetric states and only three antisymmetric ones, the symmetric states are the ortho form.

b. At high temperatures the ratio ortho/para is 6/3 = 2/1.

c. As $T \to 0$, the $J = 0$ state becomes the stable one; this is a symmetric state, and so is ortho.

d. The chance that any given molecule is ortho and so will go to the $J = 0$ level is 2/3, and this level is 6-fold degenerate; thus the probability of each of these 6 states is (1/6)(2/3) = 1/9. The probability that any one molecule is para is 1/3; then it will go to the $J = 1$ level, and this level is 9-fold degenerate (3 spin states x 3 rotation states). Thus the probability of its reaching any one of these 9 states is (1/3)(1/9) = 1/27. From these probabilities we see that the residual molar entropy is

$$\Delta s = -R\left[6\left(\frac{1}{9}\ln\frac{1}{9}\right) + 9\left(\frac{1}{27}\ln\frac{1}{27}\right)\right] = \frac{7}{3}R\ln 3 = 21.31 \text{ J K}^{-1}\text{ mol}^{-1}$$

6.17. From $z_{el} = 2[1 + \exp(-179K/T)]$ we find

$$u_{el} = RT^2\frac{\partial}{\partial T}\ln z_{el} = \frac{RT^2 2e^{-179K/T}}{z_{el}}\left(\frac{179K}{T^2}\right) = \frac{R(179\text{ K})}{e^{179K/T} + 1} = 212.9 \text{ J/mol}$$

when 100 K is substituted for T. The entropy is then

$$s_{el} = \frac{u}{T} + R\ln z = \frac{212.9 \text{ J/mol}}{100 \text{ K}} + R\ln[2(1 + e^{-1.79})] = 9.18 \text{ J K}^{-1}\text{mol}^{-1}$$

6.18. The Euler-MacLaurin formula is

$$\sum_{k=0}^{n} f(k) = \int_0^n f(x)\,dx + \frac{1}{2}[f(n) + f(0)] + \frac{1}{12}[f'(n) - f'(0)] -$$

$$- \frac{1}{720}[f^{(iii)}(n) - f^{(iii)}(0)] + \frac{1}{30240}[f^{(v)}(n) - f^{(v)}(0)] - \ldots$$

(In most cases this is an asymptotic expression rather than an equality, but we will ignore this distinction.) The function we need to sum is $f(J) = (2J + 1)\exp[-J(J + 1)\sigma]$, and the limits are 0 and ∞. At the upper limit $f(J)$ and all its derivatives vanish, so the formula reduces to

$$z_{rot} = \int_0^\infty f(J)\,dJ + \frac{1}{2}f(0) - \frac{1}{12}f'(0) + \frac{1}{720}f^{(iii)}(0) - \frac{1}{30240}f^{(v)}(0) + \ldots$$

The integral has been evaluated as $1/\sigma$; $f(0) = 1$; and the derivatives are found by tedious but straightforward differentiation:

$$f'(J) = [2 - (2J + 1)^2\sigma]e^{-J(J + 1)\sigma}; \quad f'(0) = 2 - \sigma$$

$$f^{(iii)}(J) = [-12\sigma + 12(2J + 1)^2\sigma^2 - (2J + 1)^4\sigma^3]e^{-J(J + 1)\sigma}$$

$$f^{(iii)}(0) = -12\sigma + 12\sigma^2 - \sigma^3$$

$$f^{(v)}(J) = [120\sigma^2 - 180(2J + 1)^2\sigma^3 + 30(2J + 1)^4\sigma^4 -$$

$$- (2J + 1)^6\sigma^5]e^{-J(J = 1)\sigma}$$

$$f^{(v)}(0) = 120\sigma^2 - 180\sigma^3 + 30\sigma^4 - \sigma^5$$

It can be seen that the seventh derivative would have a term in σ^3 but not in any lower power of σ; thus to get the expansion correct to terms in σ^2 we need go no further. Substituting the values found, retaining terms up to σ^2, and collecting like powers of σ, we find

$$z_{rot} = \frac{1}{\sigma} + \frac{1}{2} - \frac{1}{12}(2 - \sigma) + \frac{1}{720}(-12\sigma + 12\sigma^2 + \ldots) - \frac{1}{30240}(120\sigma^2 + \ldots)$$

$$= \frac{1}{\sigma} + \frac{1}{3} + \frac{1}{15}\sigma + \frac{4}{315}\sigma^2 - \ldots$$

7.1. Since S and V are extensive properties, they vary in direct proportion to the n's if composition, temperature, and pressure are constant. But U is also extensive and so varies in direct proportion not only to the n's but also to S and V. Therefore by Euler's theorem

$$S\left(\frac{\partial U}{\partial S}\right)_{V,n} + V\left(\frac{\partial U}{\partial V}\right)_{S,n} + \sum_i n_i\left(\frac{\partial U}{\partial n_i}\right) = U$$

We now evaluate the derivatives by using Eq. (7.1):

$$dU = T\,dS - p\,dV + \sum_i \mu_i dn_i$$

and this leads to

$$U = TS - pV + \sum_i \mu_i n_i$$

7.2. Differentiating the last equation in Problem 7.1 gives

$$dU = T\,dS + S\,dT - p\,dV - V\,dp + \sum_i(\mu_i\,dn_i + n_i\,d\mu_i)$$

Now subtracting Eq. (7.1) gives

$$0 = S\,dT - V\,dp + \sum_i n_i\,d\mu_i$$

7.3. From $Y = (n_1 + n_2)\bar{y}$ we get by differentiation

$$y_1 = \left(\frac{\partial Y}{\partial n_1}\right)_{n_2} = \bar{y} + (n_1 + n_2)\left(\frac{\partial \bar{y}}{\partial n_1}\right)_{n_2}$$

Since Y is extensive, the definition of $\bar{y}$ shows that it is intensive; that is, it can be expressed as a function of the mole fractions x_1 and x_2. Thus

$$\left(\frac{\partial \bar{y}}{\partial n_1}\right)_{n_2} = \left(\frac{\partial \bar{y}}{\partial x_1}\right)\left(\frac{\partial x_1}{\partial n_1}\right)_{n_2} + \left(\frac{\partial \bar{y}}{\partial x_2}\right)\left(\frac{\partial x_2}{\partial n_1}\right)_{n_2}$$

Now from $x_1 = n_1/(n_1 + n_2) = 1 - n_2/(n_1 + n_2)$ we find

$$\left(\frac{\partial x_1}{\partial n_1}\right)_{n_2} = \frac{n_2}{(n_1 + n_2)^2} = \frac{x_2}{n_1 + n_2}$$

Similarly we find

$$\left(\frac{\partial x_2}{\partial n_1}\right)_{n_2} = -\frac{x_2}{n_1 + n_2}$$

Substituting these values gives

$$y_1 = \bar{y} + x_2\left[\frac{\partial \bar{y}}{\partial x_1} - \frac{\partial \bar{y}}{\partial x_2}\right]$$

If $\bar{y}$ is expressed as a function of x_1 only, then $\delta\bar{y}/\partial x_2$ vanished, and vice versa.

7.4. The migration of A (in amount n) from phase α to form one mole of A_n in phase β can be represented by

$$nA^{(\alpha)} \longrightarrow A_n^{(\beta)}$$

Therefore $\Delta G^{(\alpha)} = -n\mu(A)^{(\alpha)}$ and $\Delta G^{\beta)} = \mu(A_n)^{(\beta)}$

from which $\Delta G = \mu(A_n)^{(\beta)} - n\mu(A)^{(\alpha)} \leq 0$ (1)

If, however, we ignore the association in phase β, we would find

$$\mu(A_n)^{(\beta)} = \frac{\partial G}{\partial n(A_n)} = \frac{\partial G}{\partial n(A)}\frac{\partial n(A)}{\partial n(A_n)} = n\frac{\partial G}{\partial n(A)} = n\mu(A)^{(\beta)}$$

and this substituted into (1) leads to $\mu(A)^{(\beta)} \leq \mu(A)^{(\alpha)}$.

7.5. For the solute $\mu^{sol} = \mu^{\Delta} + RT \ln a_2$

while for the vapor $\mu^{vap} = \mu^{\bullet} + RT \ln \frac{f_2}{p^{\bullet}}$

Since equilibrium requires that $\mu^{sol} = \mu^{vap}$, we find that

$$RT \ln \frac{f_2}{a_2 p^{\bullet}} = \mu^{\Delta} - \mu^{\bullet}$$

The qunatity on the right is constant, so f_2/a_2 is constant; call it K. Then $K = \frac{f_2}{a_2} = \frac{P_2}{\gamma_2 x_2}\frac{f_2}{P_2}$. Finally, since both γ_2 and $f_2/P_2 \longrightarrow 1$ as $x_2 \longrightarrow 0$, we get $K = \lim_{x_2 \to 0}\left(\frac{P_2}{x_2}\right)$.

7.6. For equilibrium we must have $\mu_2^{\alpha} = \mu_2^{\beta}$; thus

$$\mu_2^{\Delta(\alpha)} + RT \ln a_2^{(\alpha)} = \mu_2^{\Delta(\beta)} + RT \ln a_2^{(\beta)}$$

which implies that $\dfrac{a_2^{(\alpha)}}{a_2^{(\beta)}}$ is constant. To evaluate this constant, we note that $\dfrac{a_2^{(\alpha)}}{a_2^{(\beta)}} = \dfrac{x_2^{(\alpha)}\gamma_2^{(\alpha)}}{x_2^{(\beta)}\gamma_2^{(\beta)}}$; and since both the γ's $\rightarrow$ 1 as the the corresponding x's $\rightarrow$ 0, we find $\dfrac{a_2^{(\alpha)}}{a_2^{(\beta)}} = \lim \left(\dfrac{x_2^{(\alpha)}}{x_2^{(\beta)}}\right)$.

7.7. For an infinitely dilute reference state

$\mu_2^{c} = \mu_2^{\Delta} + RT \ln a_2^{c}$, where $\mu_2^{\Delta} = \lim\limits_{x_2 \to 0} \left(\mu_2 - RT \ln x_2\right)$. Therefore

$$\mu_2^{c} = \lim_{x_2 \to 0} \left(\mu_2 - RT \ln x_2\right) + RT \ln a_2^{c}$$

or

$$RT \ln a_2^{c} = \lim_{x_2 \to 0} \left(\mu_2^{c} - \mu_2 + RT \ln x_2\right)$$

and it readily follows that $a_2^{c} = \lim\limits_{x_2 \to 0} \left[x_2 \exp\left(\dfrac{\mu_2^{c} - \mu_2}{RT}\right)\right]$.

7.8. We start by making up a table:

x_2	γ_1	$\ln \gamma_1/x_2^2$	x_2	γ_1	$\ln \gamma_1/x_2^2$
0.01	0.9994	-6.002	0.06	0.9815	-5.187
0.02	0.9976	-6.007	0.07	0.9732	-5.544
0.03	0.9952	-5.346	0.08	0.9638	-5.761
0.04	0.9920	-5.020	0.09	0.9533	-5.904
0.05	0.9877	-4.951	0.10	0.9429	-5.880

For $x_2 = 0$ the extrapolated value is about -6.0. Now plot $\ln \gamma_1/x_2^2$ as abscissa vs. x_2 as ordinate and get the value of the integral from the area under the curve. Alternatively, use Simpson's rule, which gives the value as -0.556. Then

$$\ln \gamma_2 = -\frac{0.90}{0.10} \ln (0.9429) - (-0.556) = 1.085$$

and so $\gamma_2 = 2.96$

7.9. Since the y's, as derivatives of the extensive property Y, are intensive - that is, homogeneous functions of degree zero - we have by Euler's theorem

$$\sum_i n_i \left(\frac{\partial n_j}{\partial n_i}\right) = 0 \text{ (for each j)} \qquad (1)$$

and since

$$dY = \sum_i \left(\frac{\partial Y}{\partial n_i}\right) dn_i = \sum_i y_i \, dn_i$$

the reciprocity relation gives $(\partial y_i/\partial n_j)_{n'} = (\partial y_j/\partial n_i)_{n'}$. Therefore

$$\sum_i n_i \, dy_i = \sum_i n_i \sum_j \left(\frac{\partial y_j}{\partial n_i}\right) dn_j = \sum_j \sum_i n_i \left(\frac{\partial y_j}{\partial n_i}\right) dn_j = 0$$

since the inner sum vanishes by (1).

7.10. $\Pi v_1 = -RT \ln a_1$. Various choices of units are possible, but they must be consistent. Using SI units, we note that $v_1 = 18.02 \times 10^{-6}$ m^3/mol. Thus

$$\Pi = -\frac{(8.3144 \text{ J K}^{-1} \text{ mol}^{-1})(298.15 \text{ K}) \ln (0.9870)}{18.02 \times 10^{-6} \text{ M}^3 \text{ mol}^{-1}} = 1.8 \times 10^6 \text{Pa}$$

$= 18$ bars

7.11. This amounts to choosing a different path for the evaluation of the integral in Eq. (7.47):

$$\int_{1,p_0}^{x_1,\, p_0 + \Pi} \left[\left(\frac{\partial \mu_1}{\partial x_1}\right)_{p,T} dx_1 + \left(\frac{\partial \mu_1}{\partial p}\right) dp\right] = 0$$

Let the path consist of varying p from p_0 to $p_0 + \Pi$ while keeping x_1 fixed at the value 1, then letting x_1 vary from 1 to its final value at constant pressure $p_0 + \Pi$. Along the first part $dx_1 = 0$ and $(\partial\mu_1/\partial p) = v_1^*$. Along the second part

$$\left(\frac{\partial\mu_1}{\partial x_1}\right)_{p,T} = RT\,\frac{\partial \ln a_1(p_0 + \Pi)}{\partial x_1}$$

Thus this part of the integral is

$$RT \ln a_1(p_0 + \Pi)\Big]_{x_1 = 1}^{x_1}$$

Now $a_1 = 1$ when $x_1 = 1$ at all pressures only if a pressure-dependent standard chemical potential is used. In this case Eq. (7.47) becomes

$$RT \ln a_1(p_0 + \Pi) = -\int_{p_0}^{p_0 + \Pi} v_1^*\, dp$$

7.12. As in Prob. 7.11, This one involves changing the path of integration. The solute must be added at the constant temperature T_f^*; this will yield the value $R \ln a_1(T_f^*)$ when Eq. (7.35) is integrated. The temperature change must then be taken at the final concentration. This requires that Δh_{fus}, the enthalpy of fusion of the solid into pure liquid, be replaced by the enthalpy of pure solid melting into a solution of the final concentration. Thus Eq. (7.36) will be replaced by

$$\ln a_1(T_f^*) = -\int_{T_f}^{T_f^*} \frac{h_1(\underline{l}) - h_1^*(s)}{RT^2}\, dT$$

7.13. Since Δh_{fus} is to be treated as constant, Eq. (7.37) can be used:

$$\ln a_1(258.65\ \text{K}) = \frac{9.89 \times 10^3\ \text{J/mol}}{8.3143\ \text{J K}^{-1}\ \text{mol}^{-1}}\left(\frac{1}{278.65\ \text{K}} - \frac{1}{258.65\ \text{K}}\right) = -0.3301$$

$$a_1 = 0.719$$

7.14. If we are to have $\ln \gamma_2 = wx_1^2/RT$, the reference state (and standard state) for component 2 must be pure 2. Then since $\ln \gamma_2 = 0$ when $x_2 = 1$, $x_1 = 0$, we get by integrating the Gibbs-Duhem equation in the form

$$d \ln \gamma_2 = -\frac{x_1}{x_2} d \ln \gamma_1$$

$$\ln \gamma_2 = -\int_{x_2=1}^{x_2} \frac{x_1}{x_2} d \ln \gamma_1 = -\int_{x_2=1}^{x_2} \frac{x_1}{x_2}\left(\frac{2wx_2}{RT} dx_2\right) = -\frac{2w}{RT}\int_{x_2=1}^{x_2} x_1 \, dx_2$$

$$= \frac{2w}{RT}\int_0^{x_1} x_1 \, dx_1 = \frac{wx_1^2}{RT} \qquad (\text{since } dx_2 = -dx_1)$$

7.15. $\frac{dv}{dx_1} = v_1 + v_2\left(\frac{dx_2}{dx_1}\right) + x_1\left(\frac{dv_1}{dx_1}\right) + x_2\left(\frac{dv_2}{dx_1}\right) = v_1 - v_2$, since the last two terms vanish, by the generalized Gibbs-Duhem equation. Therefore

$$v + x_2\left(\frac{dv}{dx_1}\right) = x_1v_1 + x_2v_2 + x_2(v_1 - v_2) = (x_1 + x_2)v_1 = v_1$$

and $$v - x_1\left(\frac{dv}{dx_1}\right) = x_1v_1 + x_2v_2 - x_1(v_1 - v_2) = (x_1 + x_2)v_2 = v_2$$

For the H_2O-H_2SO_4 solutions we need:

x_2	x_1	$v/(cm^3/mol)$
0.06319	0.93681	19.415
0.07102	0.92898	19.616
0.07941	0.92059	19.836

We need dv/dx_1 at the intermediate point $x_1 = 0.92898$. We can approximate it by taking $\Delta v/\Delta x_1$, calculated by using the first and third values. By the theorem of mean value this is exactly equal to dv/dx_1 at some intermediate value of x_1, so it it a reasonable estimate at the value 0.92898. This gives

$$\frac{dv}{dx_1} \simeq \frac{(19.836 - 19.415)\ cm^3\ mol^{-1}}{0.92059 - 0.93681} = -25.956\ cm^3/mol$$

Thus $v_1 = v + x_2\left(\frac{dv}{dx_1}\right) = [19.616 + (0.07102)(-25.956)]\ \frac{cm^3}{mol} = 17.773\ \frac{cm^3}{mol}$

and $v_2 = v - x_1\left(\frac{dv}{dx_1}\right) = [19.616 - (0.92898)(-25.956)]\ \frac{cm^3}{mol} = 43.729\ \frac{cm^3}{mol}$

This method of estimating a derivative from three points ignores the middle point. A better method is as follows: Let the points be x - b, x, and x + a, where a and b are small positive numbers. Then

$$f'(x) = \frac{b^2 f(x + a) + (a^2 - b^2)f(x) - a^2 f(x - b)}{ab(a + b)}$$

(If a = b, this reduces to the simpler method.) Applied to the present problem, this gives $dv/dx_1 = -25.937\ cm^3/mol$, $v_1 = 17.774\ cm^3/mol$, and $v_2 = 43.711\ cm^3/mol$.

7.16. The mass of one mole of mixture is $x(EtOH)M(EtOH)+x(H_2O)M(H_2O)$. For $x_{EtOH} = 0.370$ this yields 0.370(46.07 g/mol) + 0.630(18.02 g/mol) = 28.40 g/mol. Therefore

$$v = \frac{28.40\ g/mol}{0.8911\ g/cm^3} = 31.87\ cm^3/mol$$

Similarly, using 1 for H_2O and 2 for EtOH, we find

x_1	x_2	$v/(cm^3/mol)$
0.62	0.38	32.270
0.63	0.37	31.869
0.64	0.36	31.470

Then as in Problem 7.15 we estimate

$$\frac{dv}{dx_1} \simeq \frac{(31.470 - 32.270)\ cm^3/mol}{0.64 - 0.62} = -40.00\ cm^3/mol$$

and so

$$v_1 = [31.869 + 0.37(-40.00)]\ cm^3 mol = 17.07\ cm^3/mol$$

$$v_2 = [31.869 - 0.63(-40.00)]\ cm^3/mol = 57.07\ cm^3 mol$$

The molar volumes of pure ethanol and water are given by

$$\frac{46.07\ g/mol}{0.7893\ g/cm^3} = 58.37\ cm^3/mol \text{ and } \frac{18.02\ g/mol}{0.9982\ g/cm^3} = 18.05\ cm^3/mol \text{ resp.}$$

$$\text{Thus } \Delta v_{mixing} = [0.37(57.07 - 58.37) + 0.63(17.07 - 18.05)]\ cm^3/mol$$
$$= -1.10\ cm^3/mol$$

7.17. Since the entropies in the left column are equally spaced, there is no advantage in using the more complicated formula given at the end of Prob. 7.15 for estimating the derivative $(\partial \Delta H/\partial n_1)$ at constant n_2. Then using n_1 for Et_2O and n_2 for C_7H_{16}, we find at $n_1 = 0.6$ mol,

$$\underline{l}_1 = \frac{\partial \Delta H}{\partial n_1} = \frac{(582.9 - 460.1)\ J}{(0.7 - 0.5)\ mol} = 614.0\ J/mol$$

$$\underline{l}_2 = \frac{\Delta H - n_1 \underline{l}_1}{n_2} = \frac{524.6\ J - (0.6\ mol)(614.0\ J/mol)}{1\ mol} = 156.2\ J/mol$$

Similarly at $n_1 = 1.2$ mol we get

$$\underline{l}_1 = \frac{(827.7 - 762.6)\ J}{(1.3 - 1.1) mol} = 325.5\ J/mol$$

$$\underline{l}_2 = \frac{796.8\ J - (1.2\ mol)(325.5\ J/mol)}{1\ mol} = 406.2\ J/mol$$

7.18. The problem is to maximize

$$\hat{s} = -k \sum_{N,i} \gamma(N,i) \ln \gamma(N,i)$$

subject to the conditions

$$\sum_{N,i} \gamma(N,i) = 1$$

$$\sum_{N,i} \varepsilon(N,i)\ \gamma(N,i) = \bar{u}$$

$$\sum_{N,i} N\ \gamma(N,i) = \bar{N}$$

where $\bar{N}$ is the average number of particles per system. In differential form (with the indices N and i omitted from the summation signs) these are

$$\sum \ln \gamma(N,i)\ d\gamma(N,i) = 0$$

$$\sum d\gamma(N,i) = 0$$

$$\sum \varepsilon(N,i)\ d\gamma(N,i) = 0$$

$$\sum N \, d\gamma(N,i) = 0$$

Using α, β, and λ as lagrangian parameters, we find

$$\ln \gamma(N,i) + \alpha + \beta\varepsilon(N,i) + \lambda N = 0$$

$$\gamma(N,i) = e^{-\alpha - \beta\varepsilon(N,i) - \lambda N}$$

As in the treatment of the canonical ensemble, we can eliminate α by summing $\gamma(N,i)$ to 1:

$$e^{-\alpha} = \frac{1}{\Sigma e^{-\beta\varepsilon(N,i) - \lambda N}} = \frac{1}{\Xi}$$

and so $$\gamma(N,i) = \frac{e^{-\beta\varepsilon(N,i) - \lambda N}}{\Xi}$$

$$\frac{\partial \ln \Xi}{\partial\beta} = \frac{1}{\Xi}\frac{\partial\Xi}{\partial\beta} = \frac{-\Sigma\varepsilon(N,i)e^{-\beta\varepsilon(N,i) - \lambda N}}{\Xi} = -\Sigma\varepsilon(N,i)\gamma(N,i) = -\hat{u} \quad (1)$$

$$\hat{s} = -k\Sigma\gamma(N,i)[-\beta\varepsilon(N,i) - \lambda N - \ln \Xi] = k\beta\hat{u} + k\lambda\hat{N} + k \ln \Xi \quad (2)$$

Differentiation at constant N and λ gives, with the help of (1),

$$d\hat{s} = k\beta \, d\hat{u} + k\hat{u} \, d\beta + k\frac{\partial \ln \Xi}{\delta\beta} d\beta = k\beta \, d\hat{u}$$

whence $\dfrac{d\hat{u}}{d\hat{s}} = T = \dfrac{1}{k\beta}$, as with the canonical ensemble.

Thus (2) becomes $$\hat{s} = \frac{\hat{u}}{T} + k\lambda N + k \ln \Xi$$

or $$\hat{u} = T\hat{s} - kT\lambda N - kT \ln \Xi$$

Finally, comparison with $U = TS - pV + \mu N$ gives

$$kT\lambda = -\mu \text{ and } kT \ln \Xi = pV$$

7.19. $\Delta g^E = \Delta g - RT(x_1 \ln x_1 + x_2 \ln x_2)$ and so $\Delta g = \Delta g^E + RT(x_1 \ln x_1 + x_2 \ln x_2)$. Since $n_1 + n_2 = 1$, Δg is $\overline{\Delta g}$; that is, it corresponds to $\bar{y}$ in Prob. 7.3. We now make up a table, using the second method in Prob 7.15 to estimate the derivatives:

x_2	$x_1 = x(C_8H_{18})$	$\Delta g/(\mathrm{J/mol})$	$d\Delta g/dx_1/(\mathrm{J/mol})$
0	1	0	-
0.0980	0.9020	-316.9	2020.8
0.1938	0.8062	-396.9	679.2
0.3019	0.6981	-451.3	404.8
0.4011	0.5989	-482.5	266.5
0.4941	0.5059	-503.1	

The first value of $d\Delta g/dx_1$ differs so much from the others that it appears reasonable, in the absence of other data to verify it, to reject it. From the second we get

$$\mu_1 - \mu_1^* = \Delta g + x_2 \frac{d\Delta g}{dx_1} = -396.9 \frac{\mathrm{J}}{\mathrm{mol}} + 0.1938(679.2 \frac{\mathrm{J}}{\mathrm{mol}}) = -265.3 \frac{\mathrm{J}}{\mathrm{mol}}$$

Similar calculations on the next two values give -329.1 J/mol and -375.6 J/mol respectively.

8.1. For reactants $dn(Y_i) = -y_i\, d\xi$, and for products $dn(Z_i) = z_i\, d\xi$. Therefore

$$dU = T\,dS - p\,dV + \left[\sum_i z_i \mu(Z_i) - \sum_j y_i \mu(Y_i) \right] d\xi$$

$$= T\,dS - p\,dV - Af\,d\xi$$

Since T dS and p dV vanish when entropy and pressure are constant, the first form $Af = -(\partial U/\partial \xi)_{S,V}$ follows. The others are similarly derived from the equations

$$dH = T\,dS + V\,dp - Af\,d\xi$$

$$dA = -S\,dT - p\,dV - Af\,d\xi$$

and

$$dG = -S\,dT + V\,dp - Af\,d\xi$$

8.2. For this reaction $K_p = p/p^{\circ}$; then from Eq. (8.9)

$$\ln(72.52) - \ln(26.55) = \frac{\Delta h^{\circ}}{R} \left[\frac{1}{463.1\ \mathrm{K}} - \frac{1}{485.0\ \mathrm{K}} \right]$$

whence we get $\Delta h^{\circ} = 85682$ J/uor. Since this is assumed constant, we can apply it at both temperatures to find

at 463.1 K: $Af^\ominus = RT \ln K_p = R(463.1\text{ K}) \ln (26.55/10^3) = -13972$ J/uor

$$\Delta g^\ominus = -Af^\ominus = 13972 \text{ J/uor}$$

and $$\Delta s^\ominus = \frac{\Delta h^\ominus - \Delta g^\ominus}{T} = \frac{(85682 - 13972)\text{ J/uor}}{463.1\text{ K}} = 154.8\text{ J K}^{-1}\text{uor}^{-1}$$

at 485.0 K: $Af^\ominus = R(485.0) \ln (72.52/10^3) = -10581$ J/uor $= -\Delta g^\ominus$

$$\Delta s^\ominus = \frac{(85682 - 10581)\text{ J/uor}}{485.0\text{ K}} = 154.8\text{ J K}^{-1}\text{ uor}^{-1}$$

Note that constancy of Δh implies constancy of Δs, since both result from a zero value of Δc_p. If the atmosphere were chosen as the standard pressure, $\Delta h^\ominus$ would be unchanged, since it depends only on the ratio of the two pressures. For the other properties we would have

$$Af^\ominus = -\Delta g^\ominus = RT \ln \frac{p}{p^\ominus} = R(485.0\text{ K}) \ln \frac{72.75\text{ mbar}}{1013.25\text{ mbar}} = -10621 \text{ J/uor}$$

and

$$\Delta s^\ominus = (85682 - 10621)\text{ J/uor}/(485.0\text{ K}) = 154.8\text{ J K}^{-1}\text{ uor}^{-1} \text{ at } 485.0\text{ K.}$$

$$Af^\ominus = -\Delta g^\ominus = R(463.1\text{ K}) \ln \frac{26.55\text{ mbar}}{1013.25\text{ mbar}} = -14023\text{ J/uor}$$

and $\Delta s^\ominus$ is the same (to 4 significant figures) as at 485 K.

8.3. Since Δc_p is taken to be zero, both $\Delta h^\ominus$ and $\Delta s^\ominus$ will be constant. We get the latter from (Note error: 86.69 kJ/mol is $\Delta g^\ominus$, not $Af^\ominus$.)

$$\Delta s^\ominus = \frac{\Delta h^\ominus - \Delta g^\ominus}{T} = \frac{(90.37 - 86.69)10^3\text{J/uor}}{298.15\text{ K}} = 12.34\text{ J K}^{-1}\text{ uor}^{-1}$$

$Af^\ominus = RT \ln K_p = T\,\Delta s^\ominus - \Delta h^\ominus$, from which $T = \dfrac{\Delta h^\ominus}{\Delta s^\ominus - R \ln K_p}$

$$\text{Thus } T = \frac{90.37 \times 10^3\text{ J uor}^{-1}}{(12.34 - 8.3143 \ln 0.01)\text{ J K}^{-1}\text{ uor}^{-1}} = 1785\text{ K}$$

8.4. Since Δc_p is assumed constant, the equation for $\Delta h^\ominus$ as a function of T will take the simple form $\Delta h^\ominus(T) = A + BT$. By the Gibbs-Helmholtz equation

$$\frac{\partial(Af^\ominus/T)}{\partial T} = \frac{\Delta h^\ominus}{T^2} = \frac{A}{T^2} + \frac{B}{T}$$

On integration from $T_0 = 298.15$ K to T, this gives

$$\frac{Af^{\ominus}(T)}{T} = \frac{Af^{\ominus}(T_0)}{T_0} + A\left(\frac{1}{T_0} - \frac{1}{T}\right) + B \ln \frac{T}{T_0} \qquad (1)$$

To get the constants we find from the table in App. 11

$Af^{\ominus}(T_0) = -8563$ J/mol (Minus signs omitted in table)

$\Delta h^{\ominus}(T_0) = 44104$ J/mol

$\Delta c_p^{\ominus} = -41.72$ J/(K mol)

Thus we find that

$$\Delta h^{\ominus}(T) = \Delta h^{\ominus}(T_0) + \Delta c_p^{\ominus}(T - T_0)$$

and so

$A = \Delta h^{\ominus}(T_0) - \Delta c_p^{\ominus} T_0 = 44104 \text{ J/mol} - (-41.72 \text{ J K}^{-1}\text{mol}^{-1})(298.15 \text{ K})$

$= 56543$ J/mol; and $B = \Delta c_p^{\ominus} = -41.72 \text{ J K}^{-1} \text{ mol}^{-1}$

To find the equilibrium point we must set $Af^{\ominus}(T) = 0$ in (1); thus

$$\frac{A}{T} - B \ln\left(\frac{T}{K}\right) = \frac{A + Af^{\ominus}(T_0)}{T_0} - B \ln\left(\frac{T_0}{K}\right)$$

If we now insert numerical values, write x for T/K, and cancel out the units, this reduces to $56543/x + 41.72 \ln x = 398.629$, or

$$x = \frac{56543}{398.629 - 41.72 \ln x}$$

We can try solving this equation by substituting an assumed value of x into the r.h.s. in order to get a better value, using the improved value to get a third one, and continuing iteratively. We can expect convergence, because the term in ln x on the right varies compara- slowly with x. Starting with 300, we find that the successive values of x are 351.93, 367.15, 371.40,...,373.03, after which there is no further change. The boiling point of water at one bar pressure is 372.77 K, so the error is 0.26 K, or about 0.07%.

The Newton-Raphson method would require fewer iterations, but each involves more work, so it is not obvious which is the easier method.

8.5. The molar concentration of an ideal gas is p/RT. Now let the reaction be

$$y_1Y_1 + y_2Y_2 + \ldots \rightarrow z_1Z_1 + z_2Z_2 + \ldots$$

Then

$$K_p = \frac{p(Z_1)^{z_1} p(Z_2)^{z_2}\ldots}{p(Y_1)^{y_1} p(Y_2)^{y_2}\ldots} = \frac{[C(Z_1)RT]^{z_1}\{C(Z_2)RT]^{z_2}\ldots}{[C(Y_1)RT]^{y_1}[C(Y_2)RT]^{y_2}\ldots} =$$

$$= \frac{C(Z_1)^{z_1} C(Z_2)^{z_2}\ldots}{C(Y_1)^{y_1} C(Y_2)^{y_2}\ldots} (RT)^{z_1 + z_2 + \ldots - y_1 - y_2 - \ldots}$$

$$= K_c (RT)^{\Delta n_g}$$

and so $\frac{d \ln K_c}{dT} = \frac{d}{dT}[\ln K_p - \Delta n_g \ln (RT)] = \frac{\Delta h^{\circ}}{RT^2} - \frac{\Delta n_g}{T} = \frac{\Delta h^{\circ} - \Delta n_g(RT)}{RT^2}$

$$= \frac{\Delta h^{\circ} - p\Delta v}{RT^2} = \frac{\Delta u^{\circ}}{RT^2}$$

8.6. a. $C' = 3$ (H_2, H_2S, and S); $P = 2$ (solid and gas); $R = 1$ (chemical equilibrium; no stoichiometric restrictions, since H_2 and S go into different phases). $\therefore C = 2$, $F = 2 + 2 - 2 = 2$. To see this physically, we can choose a temperature and then introduce H_2S until the H_2 pressure has a chosen value. But then the pressure of H_2S is fixed by the equilibrium, and so also the total pressure.

b. Since the ratio between SO_2 and O_2 is not specified, we must not assume one; the only extra restriction is the chemical equilibrium. $\therefore C' = 3$, $R = 1$, $C = 2$, $P = 1$, and $F = 2 + 2 - 1 = 3$. We can, for example, choose the temperature and the pressures of SO_2 and O_2, but then the pressure of SO_3 is fixed by the equilibrium.

c. $C' = 4$; there are two equilibria ($N_2O_4 \longleftrightarrow 2NO_2$ and $2NO_2 \longleftrightarrow 2NO + O_2$) and one stoichiometric restriction [$n(NO) = 2n(O_2)$]. Thus $R = 3$, $C = 1$, $P = 1$, and so $F = 2$. The same conclusion can be reached by the fact that we introduces only one substance into a homogeneous system.

d. That $C = 2$ can be seen from the fact that we made up a homogeneous system from only 2 ingredients. Alternatively, we have $C' = 6$ (HCOOH, EtOH, HCOOEt, H_2O, $HCOO^-$, and H_3O^+), 2 chemical equilibria (esterification and acid-base), 1 stoichiometric restriction ($n(HCOOEt) = n(H_2O) + n(H_3O^+)$), and electroneutrality. Thus $R = 4$, and

$C = 6 - 4 = 2$. Since there is only one phase, this leads to $F = 3$. We can, for example, choose the temperature and pressure, add the alcohol first, and then add acid until the acid/alcohol ratio reaches a chosen value. The other concentrations are then fully fixed. Complicating it further by considering other ions which may be present, such as ethylhydronium, would not change this conclusion.

e. If the amount of diamminesilver chloride added does not exceed its solubility, six species are present: H_2O, $Ag(NH_3)_2^+$, NH_3, Cl^-, Ag^+, and AgCl. There are two chemical equilibria;

$$Ag(NH_3)_2^+ \longleftrightarrow Ag^+ + 2NH_3$$

and

$$Ag^+ + Cl^- \longleftrightarrow AgCl$$

(Others, such as $Ag(NH_3)_2^+ + Cl^- \longleftrightarrow AgCl + 2NH_3$, are not independent of these.) There is no stoichiometric restriction, since the decomposition products of the complex ion do not stay all in one phase. Electroneutrality adds one restriction, making the total 3. Thus $C = 6 - 3 = 3$. Since there are 3 phases present, $F = 2$. This would not be changed by considering tha $AgCl_2^-$ ion, since it would add another species and another equilibrium. To illustrate, we can choose the temperature, add $Ag(NH_3)_2Cl$ until the concentration of a chosen species has reached a specified value, and then the other concentrations are fixed by chemical equilibria, and the pressure by the vapor pressure of the resulting solution.

If $Ag(NH_3)_2Cl$ is added until its solubility is exceeded, the changes from this discussion will be

1 more species: solid $Ag(NH_3)_2Cl$

1 more phase: solid $Ag(NH_3)_2Cl$

1 more chemical equilibrium: $Ag(NH_3)_2Cl\ (s) \longleftrightarrow Ag(NH_3)_2^+ + Cl^-$.

Thus C remains 3, but F is decreased to 1.

8.7. The strictly regular solution is defined by $\ln \gamma_1 = (w/RT)x_2^2$, $\ln \gamma_2 = (w/RT)x_1^2$. $\therefore \mu_1 = \mu_1^* + RT(\ln x_1 + \ln \gamma_1) = \mu_1^* + RT \ln x_1 + wx_2^2$; μ_2 is given by an analogous equation. $\therefore$

$$\bar{g} = x_1\mu_1 + x_2\mu_2 = x_1\mu_1^* + x_2\mu_2^* + RT(x_1 \ln x_1 + x_2 \ln x_2) + wx_1x_2$$

since $x_1x_2^2 + x_2x_1^2 = x_1x_2(X_1 + x_2) = x_1x_2$. Then, since $dx_1/dx_2 = -1$,

$$\frac{d\bar{g}}{dx_2} = \mu_2^* - \mu_2^* + RT(\ln x_2 - \ln x_1) + w(x_1 - x_2)$$

$$\frac{d^2\bar{g}}{dx_2^2} = RT\left(\frac{1}{x_2} + \frac{1}{x_1}\right) - 2w$$

$$\frac{d^3\bar{g}}{dx_2^3} = RT\left(-\frac{1}{x_2^2} + \frac{1}{x_1^2}\right)$$

Thus the third derivative is zero only when $x_1 = x_2 = 1/2$. Then to make the second derivative zero we must have

$$4RT - 2w = 0$$

and so $$T = \frac{w}{2R}$$

9.1. Given that

I $CaO \cdot Al_2O_3(s) + 8H^+ (aq) \rightarrow Ca^{2+}(aq) + 2Al^{3+}(aq) + 4H_2O(\underline{l})$
$\Delta h = -428.52$ kJ/uor

we need

II $CaO(s) + 2H^+(aq) \rightarrow Ca^{2+}(aq) + H_2O(\underline{l})$
$\Delta h = [-285.84 + (-543) - (-636)]$ kJ/uor $= -192.8$ kJ/uor

and

III $Al_2O_3(s) + 6H^+(aq) \rightarrow 2Al^{3+}(aq) + 3H_2O(\underline{l})$
$\Delta h = [3(-285.84) + 2(-525) - (-1669.8)]$ kJ/uor $= -237.7$ kJ/uor

Reversing I gives

IV $Ca^{2+}(aq) + 2Al^{3+}(aq) + 4H_2O \rightarrow CaO \cdot Al_2O_3 (s) + 8H^+(aq)$
$\Delta h = 428.52$ kJ/uor

Adding II, III, and IV gives

$CaO(s) + Al_2O_3(s) \rightarrow CaO \cdot Al_2O_3(s)$ $\quad \Delta h = -2$ kJ/mol

----

9.2. It is convenient to replace the millimoles with moles and the joules with kilojoules. Thus

$85.526Mg(OH)_2(s) + 0.192KOH + 171.244H^+(aq;1M) \longrightarrow$

$85.526Mg^{2+}(aq) + 0.192K^+ + 171.244H_2O(soln)$ $\Delta H = -9598.5$ kJ

In the second run 81.119 mol of the $Mg(OH)_2$ has been converted to MgO:

$81.119MgO(s) + 4.407Mg(OH)_2 + 0.192KOH + 171.244H^+(aq) \longrightarrow$

$85.526Mg^{2+}(aq) + 0.192K^+(aq) + 90.125H_2O(soln)$ $\Delta H = -12900.5$ kJ

Reversing the second and adding it to the first gives

$81.119Mg(OH)_2(s) \longrightarrow 81.119MgO(s) + 81.119H_2O(soln))$ $\Delta H = 3302$ kJ

Divide out the coefficient to get;

$Mg(OH)_2(s) \longrightarrow MgO(s) + H_2O(soln)$ $\Delta h = 40.706$ kJ/uor

It is given that

$H_2O(soln) \longrightarrow H_2O(\underline{l})$ $\Delta h = 0.0167$ kJ/uor

Adding gives

$Mg(OH)_2(s) \longrightarrow MgO(s) + H_2O(\underline{l})$ $\Delta h = 40.723$ kJ/uor

----

9.3. $\Delta h/(kJ/mol)$

I. $\alpha\text{-}MoO_3{\cdot}H_2O(s) + (27.92\ NaOH{\cdot}7774\ H_2O) \longrightarrow$

$Na_2MoO_4\ (soln) + (25.92\ NaOH{\cdot}7776\ H_2O)$ -71.17 ± 0.17

II. $H_2O + (27.92NaOH{\cdot}7774H_2O) \longrightarrow (27.92NaOH{\cdot}7775H_2O)$ Negligible

III. $MoO_3(s) + (27.92NaOH{\cdot}7775H_2O) \longrightarrow$

$Na_2MoO_4(soln) + (25.92NaOH{\cdot}7776(H_2O)$ -78.08 ± 0.35

Reverse I, change the sign of its Δh, and add II and III to it. The result is $MoO_3(s) + H_2O(\underline{l}) \longrightarrow MoO_3{\cdot}H_2O$ $\Delta h = (-6.91 \pm 0.39)$ kJ/mol

where the estimated error is $[(0.17)^2 + (0.35)^2]^{1/2}$.

9.4. We start by converting Johnson's data into Δh's for the reactions. The molar masses of $KBrO_3$, KBr, and I_2 are 167.0092, 119.011, and 253.8 g/mol respectively.

In step 1, 0.141123 g of $KBrO_3$ was used; this is 0.845001 mmol; $-\Delta H$ is calculated to be(0.99604 K)(105.053 cal/K) + 0.013 cal = 104.650 cal; thus $\Delta h = -(104.650\ cal)/(0.845001\ mmol) = -123.846$ cal/mmol = -123.846 kcal/mol.

In step 2, 0.09983 g KBr = 0.83883 mmol was used. ΔH includes three terms, one for electrical energy added, one for the 0.38 mK by which the final temperature fell short of the original, and 0.013 cal needed to evaporate water to saturate the dry helium with which the

KBr was protected during weighing and storage. Thus

ΔH = 4.069 cal + (0.38 x 10^{-3}K)(105.053 cal/K) − 0.013 cal = 4.096 cal
Thus Δh for this step was (4.096 cal)/(0.83883 mol) = 4.883 kcal/mol.

In step 3, 0.63734 g I_2 = 2.51119 mmol was used. ΔH is calculated similarly to the procedure in step 2:

ΔH = 3.351 cal + (0.21 x 10^{-3}K)(105.053 cal/K) − 0.012 cal = 3.361 cal

Thus Δh = (3.361 cal)/(2.51119 mmol) = 1.3384 kcal/mol; however, 3 moles will be used in calculations, so the Δh needed is 4.015 kcal per 3 moles of I_2.

To keep track of how much water is involved in the reaction and how much is mere solvent or diluent, the former will be designated by the formula H_2O and the latter by <u>aq</u>. When <u>aq</u> is preceded by a number, it will indicate a specific amount of solvent or diluent water; when preceded by ∞, it will mean a large but indefinite amount. (This is not standard practice but will reduce confusion in this problem.) Also, it is helpful to note that

23.692(HI·275.05<u>aq</u>) is the same as (23.692HI·6516.5<u>aq</u>)
and   17.692(HI·368.3<u>aq</u>)   is the same as (17.692HI·6516.5<u>aq</u>)

The reaction in step 1 is

A. $KBrO_3$(s) + 23.692(HI·275.05H_2O) → (KBr 3I_2 17.692HI 6516.5<u>aq</u> 3H_2O)

Think of this as a result of four steps:

B. $KBrO_3$(s) + 23.692(HI·275.05<u>aq</u>) → KBr(s) + 3I_2(s) + 3H_2O(<u>l</u>) +
(17.692HI·6516.5<u>aq</u>)

C. 3H_2O(<u>l</u>) + (17.692HI·6516.5<u>aq</u>) → (17.692HI 6516.5<u>aq</u> 3H_2O)

D. KBr(s) + (17.692HI·6516.5<u>aq</u> 3H_2O) → (KBr·17.692HI·6516.5<u>aq</u>·3H_2O)

E. 3I_2+(KBr·17.692HI·6516.5<u>aq</u>·3H_2O) → (KBr·3I_2·17.692HI·6516.5<u>aq</u>·3H_2O)

Adding these, we find that B + C + D + E → A. Johnson does not mention C as a separate step, presumably regarding its Δh as negligible. Therefore Δh(B) = Δh(A) − Δh(C) − Δh(D) − Δh(E) = [−123.846 − 4.883 − 4.015] kcal/uor = −132.744 kcal/uor. Reversing B and adding the remaining needed reactions gives

$\Delta h/(\text{kcal/uor})$

$KBr(s) + 3I_2(s) + 3H_2O(\underline{l}) + (17.692HI\cdot 6516.5\underline{aq}) \longrightarrow$

$KBrO_3(s) + (23.692HI\cdot 6516.5\underline{aq})$ 132.744

$KBrO_3(s) \longrightarrow \frac{3}{2} O_2(g) + KBr(s)$ -8.10

$3H_2(g) + \frac{3}{2} O_2(g) \longrightarrow 3H_2O(\underline{l})$ 3(-68.32) = -204.96

Adding these, we get

$3I_2(s) + 3H_2(g) + (17.692HI\cdot 6516.5\underline{aq}) \longrightarrow (23.692HI\cdot 6516.5\underline{aq})$ -80.316

The two solutions in this reaction differ not only in that the product contains 6 moles more of HI, but also in concentration. If we right the one on the left as 17.692(HI 368.3$\underline{aq}$) and that on the right as 23.692(HI 275.05$\underline{aq}$), we see that we must dilute 17.692(HI 275.05$\underline{aq}$) from the right side to the concentration on the left. That requires

$17.692(HI\cdot 275.05\underline{aq}) + 1650\underline{aq} \longrightarrow 17.692(HI\cdot 368.3\underline{aq})$ -0.212

Adding yields

$3I_2(s) + 3H_2(g) + 1650\underline{aq} \longrightarrow 6(HI\cdot 275.05\underline{aq})$ -80.528

The last step is to dilute the product to infinite dilution:

$6(HI\cdot 275.05\underline{aq}) + \infty\underline{aq} \longrightarrow 6HI(\underline{aq})$ -0.819

whence, by adding, we get

$3I_2(s) + 3H_2(g) + \infty\underline{aq} \longrightarrow 6HI(\underline{aq})$ -81.347

Δh_f for the iodide ion is 1/6 of this, or -13.558 kcal/mol; this should be rounded to -13.56. In SI units this is -56.73 kJ/mol.

10.1. $m(Na^+) = 0.001$; $m(Ca^{2+}) = 0.003$; $m(Cl^-) = 0.007$.
$\therefore I = (1/2)[(0.001)1^2 + (0.003)2^2 + (0.007)(-1)^2] = 0.010$
$m_\pm(CaCl_2) = [(0.003)(0.007)^2]^{1/3} = 0.00528$
$\ln \gamma_\pm = 1.171(2)(-1)(0.010)^{1/2} = -0.2342$
$\gamma_\pm = 0.791$; $a_\pm = 0.0042$

10.2. $\ln K_{sp} = \nu(\ln m_\pm + \ln \gamma_\pm) = \nu(\ln m_\pm + Cz_+z_-I^{1/2})$.
When I = 0 we find $\ln K_{sp} = \nu \ln m_\pm(0)$. When this is substituted in

the first equation and ν is canceled out, we find

$$\ln m_{\pm} = \ln m_{\pm}(0) - Cz_{+}z_{-}I^{1/2}$$

Now ln m and ln $m_{\pm}$ differ only by a constant; thus this can be reduced to

$$\ln m = \ln m(0) - Cz_{+}z_{-}I^{1/2}$$

Thus (since the z's are of opposite signs) the solubility increases with increasing ionic strength. This is called the secondary salt effect.

10.3. The appropriate form of the Gibbs-Duhem equation is (see p. 212)

$$n_{kg}\, d(\ln a_1) + m\nu[d(\ln\gamma_{\pm}) + d(\ln m) = 0$$

Expressing ln $\gamma_{\pm}$ by the Debye-Hückel theory, we get

$$n_{kg}\, d(\ln a_1) = -m\nu\left[-\frac{B}{2}\, m^{-1/2}\, dm + \frac{dm}{m}\right] = \left[\frac{B\nu m^{1/2}}{2} - \nu\right]dm$$

Since ln a_1 = 0 when m = 0, integration gives

$$n_{kg} \ln a_1 = \frac{B\nu m^{3/2}}{3} - \nu m$$

and so

$$\ln \gamma_1 = \ln a_1 - \ln x_1 = \frac{B\nu m^{3/2}}{3n_{kg}} - \frac{\nu m}{n_{kg}} - \ln x_1$$

but $\ln x_1 = \ln\left[\frac{n_{kg}}{n_{kg} + \nu m}\right] = -\ln\left[1 + \frac{\nu m}{n_{kg}}\right] = -\frac{\nu m}{n_{kg}} + \mathcal{O}(m^2)$

Substituting gives

$$\ln \gamma_1 = \frac{B\nu m^{3/2}}{3n_{kg}} + \mathcal{O}(m^2)$$

10.4. At m = 3.0 we can calculate the value of the integrand in Eq. (10.22) as follows:

$$x_1 = \frac{n_{kg}}{n_{kg} + m\nu} = \frac{55.49}{55.49 + (2)(3.00)} = 0.90242$$

$$\gamma_1 = \frac{a_1}{x_1} = \frac{0.89271}{0.90242} = 0.98924$$

Thus the integrand is $\frac{55.49}{2}\frac{\ln(0.989241)}{(3.00)^2} - \frac{1.171}{3}(3.00)^{-1/2} = -0.2587$

By treating the other concentrations similarly we can make up the following table:

m	integrand	m	integrand
0.1	-0.5903	1.0	-0.3456
0.2	-0.5118	1.5	-0.3091
0.3	-0.4737	2.0	-0.2862
0.5	-0.4129	2.5	-0.2706 (a_1=0.91244)
0.7	-0.3790	3.0	-0.2587

When these data are plotted, the graph shows a large slope and curvature, making extrapolation to zero concentration unreliable; data at lower concentrations are needed. The best that can be done with the present data is to estimate the limit as about -0.69; fortunately, any error in this value makes an error only 1/30 as large in the integral, if the following method is used. If data at m = 0.4 instead of 0.5 were listed, these figures would be well placed for a Simpson's rule integration in 3 ranges, 0-0.4, 0.4-1.0, and 1.0-3.0. By a graphical or numerical interpolation we can estimate the value -0.4407 at m = 0.4. With this value a Simpson's rule evaluation over these three ranges gives

$$-\int_0^m \left(\frac{n_{kg}}{2}\frac{\ln \gamma_1}{m^2} - \frac{B}{3}m^{-1/2}\right)dm \simeq \frac{0.1}{3}[0.69 + 0.4407 + 4(0.5903 + 0.4737) + 2(0.5118)] + \frac{0.3}{3}[0.4407 + 4(0.3790) + 0.3456] + \frac{0.5}{3}[0.3456 + 4(0.3091 + 0.2706) + 2(0.2862) + 0.2587] = 1.026_5$$

$$\text{Thus } \ln \gamma_1 = \ln(0.9024_2) - \frac{55.49}{2(3.00)}\ln(0.9892_4) - \frac{2}{3}(1.171)(3.00)^{1/2} + 1.026_5 = -0.328_2$$

whence $\gamma_{\pm} = 0.720$.

10.5. $\frac{RT}{F}$ at 25 °C is $\frac{(8.1343 \text{ J mol}^{-1}\text{ K}^{-1})(298.15\text{ K})}{96486\text{ C mol}^{-1}} = 0.02569\ \frac{\text{J}}{\text{C}} =$

$= 0.02569$ V

10.6. $\mathcal{E}^0$ for $CuSO_4(aq)|Cu$ is defined as E^0 for the cell

$$Pt,H_2|H^+(aq)||CuSO_4(aq)|Cu,Pt$$

The cell reaction is $H_2 + Cu^{2+}(aq) \longrightarrow 2H^+(aq) + Cu$, and so the emf is

$$E = \mathcal{E}^0{}_{bar} - \frac{RT}{2F}\ln\frac{a(H^+)^2}{\frac{f(H_2)}{1\ bar}\,a(Cu^{2+})} = \mathcal{E}^0{}_{atm} - \frac{RT}{2F}\ln\frac{a(H^+)^2}{\frac{f(H_2)}{1\ atm}\,a(Cu^{2+})}$$

$$\therefore \quad \mathcal{E}^0{}_{bar} - \mathcal{E}^0{}_{atm} = \frac{RT}{2F}\ln\frac{f(H_2)/(1\ atm)}{f(H_2)/(1\ bar)} = \frac{RT}{2F}\ln\frac{1\ bar}{1\ atm} =$$

$$= \frac{0.02569\ V}{2}\ln\frac{1}{1.01325} = -0.169\ mV$$

For $Cl^-(aq)|Cl_2(g),Pt$ the cell is $Pt,H_2(g)|HCl(aq)|Cl_2(g),Pt$, and its reaction is $H_2 + Cl_2 \longrightarrow 2H^+ + 2Cl^+$. It follows that

$$E = \mathcal{E}^0{}_{bar} - \frac{RT}{2F}\ln\frac{a_\pm{}^4}{\frac{f(H_2)}{1\ bar}\,\frac{f(Cl_2)}{1\ bar}} = \mathcal{E}^0{}_{atm} - \frac{RT}{2F}\ln\frac{a_\pm{}^4}{\frac{f(H_2)}{1\ atm}\,\frac{f(Cl_2)}{1atm}}$$

and so

$$\mathcal{E}^0{}_{bar} - \mathcal{E}^0{}_{atm} = \frac{RT}{2F}\ln\frac{(1\ bar)^2}{(1\ atm)^2} = (-0.02569V)\ln(1.01325) = -0.338\ mV$$

10.7a. One possibility is to use the cell $Hg(\underline{l}),Hg_2SO_4|CuSO_4(aq;m)|Cu$. The reaction is

$$2Hg(\underline{l}) + Cu^{2+}(aq) + SO_4{}^{2-}(aq) \longrightarrow Hg_2SO_4(s) + Cu(s)$$

and so

$$E = E^0 - \frac{RT}{2F}\ln\frac{1}{a(Cu^{2+})a(SO_4{}^{2-})} = E^0 + \frac{RT}{F}\ln a_\pm = E^0 + \frac{RT}{F}(\ln m + \ln \gamma_\pm)$$

Thus $E^0 = \lim_{m\to 0}(E - \frac{RT}{F}\ln m)$. Measure E for various values of m, and get E^0 by extrapolating the quantity on the right to m = 0. Then get $a_\pm$ from $a_\pm = \exp[F(E - E^0)/RT]$.

b. A possible cell is $Ca(Hg)|CaCl_2(aq)|AgCl(s), Ag(s)$, but this requires knowing the activity of calcium in the amalgam. This can be avoided by using the double cell

$$Ag, AgCl | CaCl_2(aq;m_1) | Ca(\text{in Hg}) | CaCl_2(aq;m_2) | AgCl, Ag$$

The reaction in the left half is

$$Ag(s) + CaCl_2(aq;m_1) \longrightarrow AgCl(s) + Ca(\text{in Hg})$$

In the right half it is the reverse of this, except for the different concentration of $CaCl_2$. Thus the net reaction is

$$CaCl_2(aq;m_1) \longrightarrow CaCl_2(aq;m_2)$$

and the emf is given by

$$E = \frac{RT}{2F} \ln \frac{a_\pm(1)^3}{a_\pm(2)^3} = \frac{3RT}{2F} \ln \frac{a_\pm(1)}{\gamma_\pm(2) m_\pm(2)}$$

or $$E + \frac{3RT}{2F} [\ln \gamma_\pm(2) + \ln (4^{1/3} m_2)] = \frac{3RT}{2F} \ln a_\pm(1)$$

and from this we find $\ln a_\pm(1) = \frac{1}{3} \ln 4 + \lim_{m_2 \to 0} \left(\frac{2FE}{3RT} + \ln m_2\right)$

Thus we must keep m_1 constant while measuring E at various values of m_2, and the limit required by the equation above is evaluated by extrapolation of the expression in parentheses.

c. This is similar to b. The net reaction (with m_1 in the left half, m_2 in the right) is

$$NaOH(aq;m_1) \longrightarrow NaOH(aq;m_2)$$

and the emf is $E = \frac{RT}{F} \ln \frac{a_\pm(1)^2}{a_\pm(2)^2} = \frac{2RT}{F}[\ln a_\pm(1) - \ln m_2 - \ln \gamma_\pm(2)]$

whence $\ln a_\pm(1) = \lim_{m_2 \to 0} \left(\frac{FE}{2RT} + \ln m_2\right)$.

10.8. Defining two quantities J and K by

$$J = E + \frac{RT}{F} \ln m_\pm^{2} = E + (0.05138\ V) \ln m$$

and $$K = E + \frac{2RT}{F}\left(\ln m - Bm^{1/2}\right) = J - (0.06016\ V) m^{1/2}$$

We then need to plot J vs. $m^{1/2}$ and K vs. m. The values to be plotted are given in the following table:

m	K/V	$m^{1/2}$	J/V
0.005	0.22179	0.07071	0.22604
0.006	0.22171	0.07746	0.22637
0.007	0.22165	0.08367	0.22667
0.008	0.22155	0.08944	0.22693
0.009	0.22147	0.09846	0.22717
0.010	0.22139	0.1000	0.22741

The graph of K vs. m gives an excellent straight line, except that the point at m = 0.007 is slightly off. Extrapolation to m = 0 (ignoring the deviant point) gives 0.22219 V, in agreement with the accepted value of 0.2222 V; linear regression gives the same result. The graph of J vs $m^{1/2}$ shows significant curvature, and extrapolation to m = 0 is unreliable.

10.9. The emf is given (with the constant molality 0.1 designated by m(2)) by by

$$E = \frac{RT}{F}\ln\frac{a_{\pm}(1)^2}{a_{\pm}(2)^2} = \frac{2RT}{F}\ln\frac{m(1)\gamma_{\pm}(1)}{a_{\pm}(2)} \qquad (1)$$

from which

$$\ln a_{\pm}(2) = \ln m(1) + \ln \gamma_{\pm}(1) - \frac{FE}{2RT} = \ln m(1) - \frac{FE}{2RT} - Bm^{1/2} + B'm$$

where B′ is defined in Eq. (10.27). We thus need to plot

$\ln m(1) - \frac{FE}{2RT} - Bm^{1/2}$ vs. m and extrapolate to zero to get $\ln a_{\pm}(2)$.

The data to be plotted are

m(1)	$\ln m(1) - 19.461E/V - 1.171m(1)^{1/2}$
0.2	-2.7707
0.3	-2.8577
0.5	-3.0113
0.7	-3.1515
1.0	-3.3434

By linear regression the limit is given by -2.6417; however, the graph appears slightly curved, and a value as high as -2.59 may be justifiable. Using the linear regression limit, we find at 0.1m

$$a_{\pm} = e^{-2.6417} = 0.0712$$

and

$$\gamma_{\pm} = a_{\pm}/m = 0.712$$

At m(1) = 1, (1) gives

$$\frac{FE}{2RT} = (19.641)(0.11165) = \ln \frac{a_{\pm}(1)}{0.0712}$$

from which $a_{\pm}(1) = 0.638 = \gamma_{\pm}(1)$.

10.10. If the reaction is written as

$$\frac{1}{2} H_2(g) + AgCl(s) \longrightarrow Ag(s) + H^+(aq) + Cl^-(aq)$$

the number of electrons involved is 1, and

$$Af^0 = FE^0 = [21.741 - 5.93\text{x}10^{-2}(t - 20) - 1.16\text{x}10^{-4}(t - 20)^2]\text{kJ/uor}$$

Since dt/dT = 1/K (t is regarded as dimensionless), the entropy is

$$\Delta s^0 = F(dE^0/dT) = F(dE^0/dt)/K = F[-6.15\text{x}10^{-4} - 2.4\text{x}10^{-6}(t - 20)]\text{V/K}$$

$$= [-59.3 - 0.23(t - 20)] \text{ J K}^{-1} \text{ our}^{-1}$$

$$\Delta h^0 = FT^2 \frac{d(E^0/T)}{dT} = F\left(T \frac{dE^0}{dT} - E^0\right) = F\left(\frac{T}{K} \frac{dE^0}{dt} - E^0\right)$$

$$= F\left(\frac{T}{K}[-6.15\text{x}10^{-4} - 2.4\text{x}10^{-6}(t - 20)] - 0.22533 + 6.15\text{x}10^{-4}(t - 20) + 1.2\text{x}10^{-6}(t - 20)^2\right)\text{V}$$

Note the units: since F is in Coulombs/mol, the volts symbol at the end is needed to convert it to J/mol, or J/uor in this case. At this point it is expedient to eliminate either t or T. Choosing to retain T, we replace t - 20 with T/K - 293.15. When we make this substitution and collect like powers of T, we find

$$\Delta h^0 = F\left[-0.30249 - 1.2\text{x}10^{-6}\left(\frac{T}{K}\right)^2\right]\text{V} = \left[-29.186 - 1.158\text{x}10^{-4}\left(\frac{T}{K}\right)^2\right] \text{ kJ/uor}$$

and

$$\Delta c_p{}^0 = \frac{d\Delta h^0}{dT} = -0.2316\left(\frac{T}{K}\right) \text{ J K}^{-1} \text{ uor}^{-1}$$

10.11. In the left half cell the reaction is

$$MgO(s) + 2F^- \longrightarrow MgF_2(s) + (1/2)O_2(g) +$$

and in the right

$$MgF_2(s) + (1/2)O_2 + 2\varepsilon^- \longrightarrow MgO_2(\text{solid soln.}) + 2F^-$$

Thus the net reaction is

$$MgO(s) \longrightarrow MgO(\text{solid soln.}) \quad (z = 2)$$

Therefore, since the activity of the pure solid MgO is 1,

$$E = \frac{RT}{2F} \ln \frac{a(MgO;\ s)}{a(MgO;\ s.\ sln.)} = -\frac{RT}{2F} \ln a(MgO;\ s.\ sln.)$$

Thus, for example, at 1163 K and x(MgO) = 0.2, we find

$$\ln a(MgO;\ x = 0.2) = -\frac{2FE}{RT} = -\frac{2(96487)(51.01\text{x}10^{-3})}{(8.13143)(1163)} = -1.018$$

$$a(MgO; x = 0.2) = 0.3613$$

In this manner we can set up the following table (using subscript 1 for MgO, 2 for MnO):

x_1	a_1 at 1163 K	a_1 at 1223 K	a_1 at 1318 K
0.2	0.3613	0.3488	0.2956
0.3	0.5374	0.5002	0.4598
0.4	0.5397	0.5346	0.5247
0.5	0.6446	0.6451	0.6382
0.7	0.7578	0.7667	0.7611
0.9	0.8978	0.8839	0.8879
(1.0)	(1.0)	(1.0)	(1.0)

The most appropriate form of the Gibbs-Duhem equation is

$$d \ln a_2 = -\frac{x_1}{x_2} d \ln a_1$$

If we choose convention I ($a_2 = 1$ when $x_1 = 0$), this integrates to

$$\ln a_2 = -\int_{x_1=0}^{x_1} \frac{x_1}{x_2}\, d\ln a_1 = -\int_{x_1=0}^{x_1} \frac{x_1}{x_2 a_1}\, da_1$$

The integrand remains finite as $x_1 \to 0$; we can get its value by a suitable extrapolation. The data needed are

x_1	a_1	$x_1/(x_2a_1)$
0.2	0.3488	0.7167
0.3	0.5002	0.8568
0.4	0.5346	1.2470
0.5	0.6451	1.5501
0.7	0.7667	3.0433

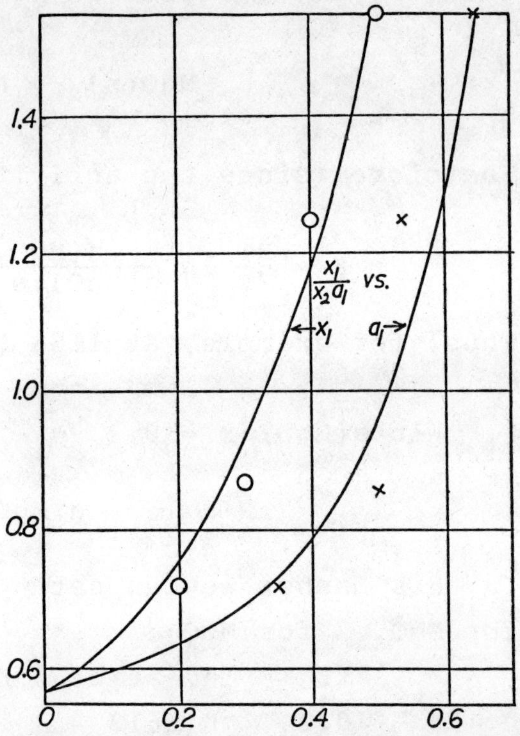

Graphical extrapolation is not very reliable, but appears to give a value of about 0.57. From a graph of $x_1/(x_2a_1)$ vs. a_1 (see figure), we can read the following values:

a_1	$x_1/(x_2a_1)$
0	0.57
0.1	0.60
0.2	0.64
0.3	0.70
0.4	0.79
0.5	0.96
0.6	1.28
0.6451	1.55

Integration by Simpson's rule from 0 to 0.6 gives 0.458, by the trapezoidal rule from 0.6 to 0.6451 gives 0.064; the total integral is $0.522 = -\ln a_2$. Thus $a_2 = 0.593$.

At $x_1 = x_2 = 0.5$ and 1223 K

$$g_{real} = x_1\mu_1^* + x_2\mu_2^* + RT(x_1 \ln a_1 + x_2 \ln a_2)$$

$$g_{ideal} = x_1\mu_1^* + x_2\mu_2^* + RT(x_1 \ln x_1 + x_2 \ln x_2)$$

Thus $$g_{real} - g_{ideal} = RT\left(x_1 \ln \frac{a_1}{x_1} + x_2 \ln \frac{a_2}{x_2}\right)$$

$$= RT\left(0.5 \ln \frac{0.645}{0.5} + 0.5 \ln \frac{0.593}{0.5}\right) = 2.16 \text{ kJ/mol}$$

By Eq. (7.30) $$h_1 - h_1^* = -RT^2 \left(\frac{\partial \ln a_1}{\partial T}\right)$$

Using the formula given in Prob. 7.15, we find

$$\frac{\partial \ln a_1}{\partial T} = \frac{60^2 \ln(0.2956) + (95^2 - 60^2) \ln(0.3488) - 95^2 \ln(0.3613)}{(60)(95)(155)}$$

$$= -1.034\text{x}10^{-3} \text{ K}^{-1}$$

$$\therefore h_1 - h_1^* = (8.3143 \frac{\text{J}}{\text{K mol}})(1223 \text{ K})^2(-1.034\text{x}10^{-3} \text{ K}^{-1}) = -12.86 \text{ kJ/mol}$$

10.12. Across the left Pt|sln.1 junction the change is

$$\frac{1}{2} H_2 \longrightarrow H^+ + \varepsilon^- \qquad (1)$$

Across the sln.1|sln.2 junction the changes are

$$t_+H^+(\text{sln.1}) \longrightarrow t_+H^+(\text{sln. 2}) \qquad (2)$$

and

$$t_-A^-(\text{sln. 2}) \longrightarrow t_-A^-(\text{sln. 1}) \qquad (3)$$

From (1), (2), and (3) we find that the net change in sln. 1 is

$$+t_-[H^+(\text{sln. 1}) + A^-(\text{sln. 1})] \qquad (4)$$

Similarly across the sln. 2|sln. 3 junction the changes are

$$t'_+H^+(\text{sln. 2}) \longrightarrow t'_+H^+(\text{sln 3}) \qquad (5)$$

and

$$t'_-A^-(\text{sln. 3}) \longrightarrow t'_-A^-(\text{sln. 2}) \qquad (6)$$

From (2), (3), (5), and (6) we find the net change in sln. 2 to be

$$t_+H^+ - t_-A^- - t'_+H^+ + t'_-A^- = +(t'_- - t_-)[H^+(2) + A^-(2)] \qquad (7)$$

since $$t_+ - t'_+ = [(1 - t_-) - (1 - t'_-)] = t'_- - t_-$$

The treatment of the changes in sln. 3 is similar to that in sln. 1 (but reversed) and leads to the conclusion that the net change is

$$-t'_-[H^+(3) + A^-(3)] \qquad (8)$$

Combining (4), (7), and (8), we find

$$-Af = \Delta g = t_-\mu_1 + (t'_- - t_-)\mu_2 - t'_-\mu_3 = t_-(\mu_1 - \mu_2) + t'_-(\mu_2 - \mu_3)$$

and so

$$E = \frac{1}{F}\,[t_-(\mu_2 - \mu_1) + t'_-(\mu_3 - \mu_2)]$$

Extending this to a cell with many liquid junctions, we find

$$E = \frac{1}{F}\,[t_-(\mu_2 - \mu_1) + t'_-(\mu_3 - \mu_2) + t''_-(\mu_4 - \mu_3) + \ldots]$$

This is the type of sum whose limit defines an integral, so in the limit of a continuously varying concentration, we have

$$E = \frac{1}{F}\int_{\mu_1}^{\mu_2} t_-\,d\mu = \frac{2RT}{F}\int_{a_\pm=a_\pm(1)}^{a_\pm(2)} t_-\,d(\ln a_\pm)$$

10.13. The reaction is

$$Na_2WO_4{\cdot}2H_2O(s) + aq \rightarrow 2Na^+(aq) + WO_4^{2-}\ (aq) + 2H_2O$$

and so the equilibrium constant is

$$K_{eq} = a_+{}^2a_-a(H_2O)^2 \quad \text{or} \quad \ln K_{eq} = 3RT\ \ln a_\pm + 2RT\ \ln a(H_2O)$$

Thus

$$-\Delta g^0 = 680\ \text{J/mol} = 3RT(\ln m_\pm + \ln \gamma_\pm) + 2RT\ \ln a(H_2O)$$

$$= 3RT(\tfrac{1}{3}\ \ln 4 + \ln m + \ln \gamma_\pm) + 2RT\ \ln\ (0.875)$$

This reduces to $\ln m = -0.2816 - \ln \gamma_\pm$. To get $\ln \gamma_\pm$ from the formula given by Dellien we need to recognize that for a 2-1 type of electrolyte, $I = 3m$. Starting with $m = 2$, $I = 6$, we get

$$\ln \gamma_\pm = -\frac{(2.3444)6^{1/2}}{1 + (1.2407)6^{1/2}} + (0.0171)6 + (0.000221)6^{3/2} = -1.316$$

$\ln m = -0.2816 - (-1.316) = 1.034$, from which $m = 2.813$. Using thie estimate of m in a similar calculation, we find $\ln \gamma_\pm = -1.329$, $m = 2.852$. Further iterations give the same result, showing that this answer is accurate enough.

CHAPTER 11

11.1. The surface tension acts on both the inside and outside surfaces. Since the thickness is negligible, both have the radius r and excess pressure $2\gamma/r$; the total excess pressure is therefore $4\gamma/r$.

11.2. Because the radius is small compared to the depth, the pressure on all parts of the bubble is practically constant, leading to a constant radius of curvature and so to a spherical surface. Soon after the formation of a bubble begins, the surface is nearly flat, the center of curvature is inside the tube, and so the radius of curvature is greater than r. When the bubble becomes hemispherical, its center of curvature lies in the plane of the tube opening, and the radius is r. As the bubble grows larger, the radius again becomes greater than r. Thus r is the minimum value of the radius, and so the maximum pressure exerted by surface tension is $2\gamma/r$. Added to the hydrostatic pressure p_0, this gives the minimum pressure needed to maintain flow.

Incidentally, whether the bubble forms on the inner or outer radius of the tube depends on surface tension and how strongly the liquid wets the material of the tube.

11.3. Eq. (11.16) becomes, on dropping the term in dT,

$$d\gamma + \sum_{i=1}^{C} \Gamma_i \, d\mu_i = 0 \qquad (1)$$

The Gibbs-Duhem equation is

$$\sum_{i=1}^{C} x_i \, d\mu_i = 0$$

Solving for $d\mu_1$ gives

$$d\mu_1 = -\sum_{i=2}^{C} \frac{x_i}{x_1} \, d\mu_i$$

Substituting this into (1) gives

$$-d\gamma = -\Gamma_1 \sum_{i=2}^{C} \frac{x_i}{x_1} d\mu_i + \sum_{i=2}^{C} \Gamma_i \, d\mu_i = \sum_{i=2}^{C} \left(\Gamma_1 - \frac{x_i}{x_1}\Gamma_1\right)$$

11.4. Let the boundary move to the left so as to increase the total amount of matter (expressed in moles) in A" by δN. Then the new values Γ'_1 and Γ'_2 are given by $\Gamma'_1 = \Gamma_1 + x_1\delta N$ and $\Gamma'_2 = \Gamma_2 + x_2\delta N$. Then

$$\Gamma'_{2(1)} = \Gamma'_2 - \frac{x_2}{x_1}\Gamma'_1 = \Gamma_2 + x_2\delta N - \frac{x_2}{x_1}(\Gamma_1 + x_1\delta N) = \Gamma_2 - \frac{x_2}{x_1}\Gamma_1 = \Gamma_{2(1)}$$

11.5 Eq. (11.11) is

$$\ln \frac{P}{P^0} = \frac{2v\gamma}{RTr}$$

The data are $v = 18\ \text{cm}^3/\text{mol} = 1.8\text{x}10^{-5}\ \text{m}^3/\text{mol}$; $T = 293$ K; $\gamma = 72.75\text{x}10^{-3}\ \text{J/m}^2$; $r = 10^{-8}$ m. Substituting gives

$$\ln \frac{P}{P^0} = \frac{2(1.8\text{x}10^{-5}\ \text{m}^3\text{mol}^{-1})(72.75\text{x}10^{-3}\ \text{J m}^{-3})}{(8.3143\ \text{J K}^{-1}\text{mol}^{-1})(293\ \text{K})(10^{-8}\ \text{m})} = 0.1075$$

and so $P/P^0 = e^{0.1075} = 1.11$.

11.6. Eq. (11.22) is $dp = -\rho g\, dz$. For an ideal gas $pV = nRT = (m/M)RT$, and so $\rho = m/V = pM/RT$. Thus $dp = -(pMg/RT)\, dz$. This integrates to

$$\ln \frac{p_2}{p_1} = -\frac{Mg(z_2 - z_1)}{RT}$$

For "air" $M = 0.029$ kg/mol, $z_2 - z_1 = 10^3$m, $g = 9.8\ \text{ms}^{-2}$, $p_1 = 1$ bar, and $T = 298$ K. Substitution in the above equation gives

$\ln \frac{p_2}{1\ \text{bar}} = -0.115$, $p_2 = 0.891$ bar. The figures for CO_2 are the same, except that $M = 0.044$ kg/mol and $p_1 = 325\text{x}10^{-6}$ bar. These figures give $p_2 = 2.73\text{x}10^{-4}$ bar. The concentration (mole fraction) of CO_2 is $2.73\text{x}10^{-4}/0.891 = 306$ parts per million.

CHAPTER 12

12.1. Prob. (3.3) $[(\partial S/\partial V)_T = \alpha/\kappa_T]$ by Jacobians:

$$\left(\frac{\partial S}{\partial V}\right)_T = \frac{J(T,S)}{J(T,V)} = \frac{J(p,V)}{J(T,V)} = \frac{V\alpha}{V\kappa_T} = \frac{\alpha}{\kappa_T} \quad \text{[by Eqs. (12.10) and (12.11)]}$$

$$(3.4): \left(\frac{\partial T}{\partial V}\right)_S = \frac{J(T,S)}{J(V,S)} = \frac{J(p,V)}{J(V,S)} = \frac{J(p,V)}{J(V,T)\frac{J(V,S)}{J(V,T)}} = \frac{V\alpha}{-V\kappa_T\left(\frac{\partial S}{\partial T}\right)_V} = -\frac{T\alpha}{\kappa_T C_V}$$

$$(3.5a): \left(\frac{\partial U}{\partial V}\right)_T = \frac{J(U,T)}{J(V,T)} = \frac{TJ(S,T) - pJ(V,T)}{J(V,T)} \quad \text{[by Eq. (12.6)]}$$

$$= \frac{TJ(V,p)}{J(V,T)} - p = T\left(\frac{\partial p}{\partial T}\right)_V - p$$

$$(3.5b): \left(\frac{\partial A}{\partial V}\right)_S = \frac{J(A,S)}{J(V,S)} = \frac{-SJ(T,S) - pJ(V,S)}{J(V,S)} = \frac{-SJ(p,V)}{J(V,S)} - p \quad \text{[Eq.(12.8)}$$

$$= S\left(\frac{\partial p}{\partial S}\right)_V - p$$

(3.6): For $C_p - C_v$ see p. 257.

$$V(\kappa_T - \kappa_S) = \frac{J(V,T)}{J(T,P)} - \frac{J(V,S)}{J(S,p)} = \frac{J(V,T)J(S,p) - J(V,S)J(T,p)}{J(T,p)J(S,p)}$$

$$= \frac{J(T,V)J(p,S) + J(V,S)J(p,T)}{J(p,T)J(p,S)} = \frac{J(p,V)J(T,S)}{J(p,T)J(p,S)} \quad \text{[Eq. (12.13) or (12.15)]}$$

$$= \frac{(V\alpha)^2}{C_p/T}, \text{ whence } \kappa_T - \kappa_S = \frac{TV\alpha^2}{C_p} \quad \text{[By Eqs. (12.10) \& (12.11)}$$

12.2. From Shaw's table we find

$$\left(\frac{\partial U}{\partial V}\right)_p = \frac{J(U,p)}{J(V,p)} = \frac{-Tc + pb}{-b} = T\frac{c}{b} - p \; ; \quad \left(\frac{\partial G}{\partial T}\right)_S = \frac{J(G,S)}{J(T,S)} = \frac{-Sb + Vc}{b} = V\frac{c}{b} - S$$

We can eliminate c/b by multiplying the 1st eq. by V, the 2nd by T, and subtracting. This give the required relation:

$$V\left(\frac{\partial U}{\partial V}\right)_p - T\left(\frac{\partial G}{\partial T}\right)_S = TS - pV$$

12.3. To simplify writing we will use X for $(\partial p/\partial T)_V$ and μ without the subscript JT for the Joule-Thomson coefficient. Thus we find

$$\left(\frac{\partial p}{\partial T}\right)_V = \frac{J(V,p)}{J(V,T)} = -b/a; \quad C_v = \frac{J(U,V)}{J(T,V)} = Tn/a; \quad C_p = \frac{J(H,p)}{J(T,p)} = Tc/\underline{1}; \text{ and}$$

$\mu = \dfrac{J(T,H)}{J(p,H)} = (Tb - V\underline{l})/Tc$. Writing these as linear equations gives

$$aX + b = 0 \qquad (1)$$

$$aC_v - Tn = 0 \qquad (2)$$

$$Tc - \underline{l}C_p = 0 \qquad (3)$$

$$Tb - \mu Tc - V\underline{l} = 0 \qquad (4)$$

We must treat one of the letters a, b, c, $\underline{l}$, and n as known, and eliminate the others. Choosing $\underline{l}$ for this role, we get from (3) $c = C_p\underline{l}/T$. Then (4) becomes $Tb - \mu C_p\underline{l} - V\underline{l} = 0$, whence $b = (\mu C_p + V)\underline{l}/T$. From (1) $a = -b/X = -(\mu C_p + V)\underline{l}/TX$, and lastly from (2) $n = aC_v/T = -(\mu C_p + V)C_v\underline{l}/(T^2X)$. Now these must be substituted into Eq. (12.17): $ac + b^2 - n\underline{l} = 0$. As expected, $\underline{l}^2$ appears in every term and can be canceled out. This leaves

$$\left(-\frac{\mu C_p + V}{TX}\right)\left(\frac{C_p}{T}\right) + \left(\frac{\mu C_p + V}{T}\right)^2 + \frac{(\mu C_p + V)C_v}{T^2X} = 0$$

The factor $(\mu C_p + V)/T^2$ occurs in every term; removing it leaves

$$-\frac{C_p}{X} + \mu C_p + V + \frac{C_v}{X} = 0$$

from which

$$C_p - C_v = \left(\mu_{JT}C_p + V\right)\left(\frac{\partial p}{\partial T}\right)_V$$

Since this difference is known to be $TV\alpha^2/\kappa_T$, it is of interest to show that the two expressions are equal. To do this note that the right side of the equation above can be expressed as

$$\left(\frac{J(H,T)}{J(H,p)}\,\frac{J(H,p)}{J(T,P)} + V\right)\frac{J(p,V)}{J(T,V)} = \left(-\frac{TJ(S,T) + VJ(p,T)}{J(p,T)} + V\right)\frac{\alpha}{\kappa_T} = T\frac{J(T,S)}{J(p,T)}\,\frac{\alpha}{\kappa_T}$$

$$= \frac{TV\alpha^2}{\kappa_T} \quad \text{[By Eqs. (12.2), (12.7), and (12.10)-(12.12)]}$$

12.4. (a) yields four equations; the other four duplicate these. They are

$$\left(\frac{\partial H}{\partial T}\right)_p - \left(\frac{\partial U}{\partial T}\right)_V = T\left(\frac{\partial p}{\partial T}\right)_V\left(\frac{\partial V}{\partial T}\right)_p; \quad -\left(\frac{\partial A}{\partial S}\right)_V + \left(\frac{\partial G}{\partial S}\right)_p = S\left(\frac{\partial V}{\partial S}\right)_p\left(\frac{\partial p}{\partial S}\right)_V$$

$$\left(\frac{\partial U}{\partial p}\right)_S - \left(\frac{\partial A}{\partial p}\right)_T = p\left(\frac{\partial S}{\partial p}\right)_T\left(\frac{\partial T}{\partial p}\right)_S; \qquad \left(\frac{\partial G}{\partial V}\right)_T - \left(\frac{\partial H}{\partial V}\right)_S = -V\left(\frac{\partial T}{\partial V}\right)_S\left(\frac{\partial S}{\partial V}\right)_T$$

(b) This one yields only two equations:

$$U - G = TS - pV \text{ and } H - A = pV + TS$$

(c) This one leads to a family of eight:

$$\left(\frac{\partial U}{\partial V}\right)_T = T\left(\frac{\partial p}{\partial T}\right)_V - p; \qquad -\left(\frac{\partial H}{\partial S}\right)_V = V\left(\frac{\partial T}{\partial V}\right)_S - T; \qquad \left(\frac{\partial G}{\partial p}\right)_S = -S\left(\frac{\partial V}{\partial S}\right)_p + V$$

$$\left(\frac{\partial A}{\partial T}\right)_p = p\left(\frac{\partial S}{\partial p}\right)_T - S; \qquad -\left(\frac{\partial U}{\partial S}\right)_p = p\left(\frac{\partial T}{\partial p}\right)_S - T; \qquad \left(\frac{\partial H}{\partial p}\right)_T = -T\left(\frac{\partial V}{\partial T}\right)_p + V$$

$$\left(\frac{\partial G}{\partial T}\right)_V = V\left(\frac{\partial S}{\partial V}\right)_T - S; \qquad \left(\frac{\partial A}{\partial V}\right)_S = S\left(\frac{\partial p}{\partial S}\right)_V - p$$

(d) Only one this time: $U - A = H - G$.

App. 2.1. From $z = -x^2 + 4xy - 5y^2 - 4x + 8y$ we get

$$\frac{\partial z}{\partial x} = -2x + 4y - 4 \quad \text{and} \quad \frac{\partial z}{\partial y} = 4x - 10y + 8$$

To get the highest point on the hill we set each of these equal to 0; the solution of the resulting pair of equations yields $y = 0$, $x = -2$, $z = 4$. To get the highest point on the path by the lagrangian method we set $(-2x + 4y - 4)dx + (4x - 10y + 8)dy = 0$. The restriction $y = 2x$ becomes $2dx - dy = 0$. Multilpying this by λ and adding gives $(-2x + 4y - 4 + 2\lambda)dx + (4x - 10y + 8 - \lambda)dy = 0$. After setting each of these to 0, we can eliminate λ by multiplying the coefficient of dy by 2 and adding. The result is

$$6x - 16y + 12 = 0$$

Eliminating y by using the restricting condition $y = 2x$, we can solve this to get $x = 6/13$ and $y = 12/13$. Substituting into the equation for the hill then gives $z = 468/169 = 2.769$. To get this result by substitution:

$$z = -x^2 + 4x(2x) - 5(2x)^2 - 4x + 8(2x) = -13x^2 + 12x$$

For a maximum we set the derivative $-26x + 12 = 0$, whence $x = 6/13$ as in the other method.

APPENDIX 2

App. 2.2. We must maximize xy subject to the restriction that $x^2 + 9y^2 = 36$. This requires that $x\,dy + y\,dx = 0$ while $x\,dx + 9y\,dy = 0$. Using $-\lambda$ as the lagrangian multiplier, we get

$$y - \lambda x = 0 \quad \text{and} \quad x - 9\lambda y = 0$$

Thus $x - 9(y/x)y = 0$ and so $x^2 = 9y^2$. Putting this into the equation for the ellipse, we get $18y^2 = 36$, $y^2 = 2$, $x^2 = 18$. Thus $x^2y^2 = 36$ and xy = area = 6.

App. 2.3. Intersection occurs when $p = k_S V^{-\gamma} = k_T V$. Therefore

$$\frac{k_S}{k_T} = V^{\gamma-1} \tag{1}$$

For the adiabatic curve $\tan \Theta_S = \frac{d}{dV}(k_S V^{-\gamma}) = -\gamma k_S V^{-\gamma-1}$

For the isothermal curve $\tan \Theta_T = \frac{d}{dV}(k_T V^{-1}) = -k_T V^{-2}$

$$\therefore \frac{\tan \Theta_S}{\tan \Theta_{T_T}} = \frac{\gamma k_S V^{-\gamma-1}}{k\ V^{-2}} = \frac{\gamma k_S}{k_T} V^{1-\gamma} = \gamma \qquad \text{by (1)}$$

Thus we need the extremum of $\Theta_S - \Theta_T$ subject to the condition

$$\tan \Theta_S = \gamma \tan \Theta_T \tag{2}$$

The lagrangian method then requires that

$$1 - \lambda \sec^2\Theta_S = 0 \quad \text{and} \quad 1 - \lambda\gamma \sec^2\Theta_T = 0$$

Subtraction and dividing out λ gives

$$\gamma \sec^2\Theta_T = \sec^2\Theta_S$$

or by trigonometry

$$\gamma(1 + \tan^2\Theta_T) = 1 + \tan^2\Theta_S$$

Substituting $\tan \Theta_S$ from (2) gives $\gamma(1 + \tan^2\Theta_T) = 1 + \gamma^2\tan^2\Theta_T$, from which we find $\tan^2\Theta_T = 1/\gamma$, and so by (2) $\tan^2\Theta_S = \gamma$. For $\gamma = 5/3$ this leads to $\Theta_S = 52.24°$, $\Theta_T = 37.76°$, $|\Theta_S - \Theta_T| = 14.48°$. (Actually, both Θ's are negative, but we can ignore this in calcualting the absolute value of the difference.)